Integrating Environment and Economy

What is the relationship between economic development and the environment? Can this relationship be improved through new forms of policy and practice?

In recent years governments at all levels have introduced numerous environmental policies in an attempt to respond to the negative impacts of economic development. Despite some important successes, it is generally accepted that the policies that have been introduced have not been able to address these negative impacts in a comprehensive way.

In response to the apparent limits of existing approaches to policy, governments at all levels have begun to explore the potential of new ways of furthering environmental objectives. While many initiatives have attempted to improve the design or the delivery of explicitly environmental policies, an increasing number of initiatives have also attempted to integrate environmental objectives into economic policies. Such initiatives could expand the influence of environmental policy dramatically. However, many parties concerned with economic growth and development are concerned about the economic implications of environmental protection. It is clear therefore that any discussion on strategies for integration must take economic as well as environmental interests into account.

Integrating Environment and Economy provides a detailed and accessible examination of the policies and practices associated with local and regional strategies for integrating economic development and environmental management. Using the idea of ecological modernization as a conceptual framework, this book considers how integrated approaches to policy might be developed and whether these integrated approaches enable economic and social activities to support rather than undermine the realization of environmental objectives.

Andrew Gouldson is a lecturer in environmental policy at the London School of Economics and **Peter Roberts** is Professor of European Strategic Planning at the University of Dundee.

Integrating Environment and Economy

Strategies for local and regional government

Edited by Andrew Gouldson and Peter Roberts

London and New York

First published 2000 by Routledge
11 New Fetter Lane, London EC4P 4EE

Simultaneously published in the USA and Canada
by Routledge
29 West 35th Street, New York, NY 10001

Routledge is an imprint of the Taylor & Francis Group

© 2000 Edited by Andrew Gouldson and Peter Roberts

Typeset in Galliard by Steven Gardiner Ltd, Cambridge

British Library Cataloguing in Publication Data
A catalogue record for this book is available
from the British Library

Library of Congress Cataloging in Publication Data
Integrating environment and economy : strategies for local and
    regional government / edited by Andrew Gouldson and Peter Roberts.
        p.   cm.
    Includes bibliographical references and index.
    1. Environmental economics.   2. Environmental policy—Economic
aspects.   3. Economic development—Environmental aspects.
I. Gouldson, Andrew.   II. Roberts, Peter, 1947–   .
HC79.E5I5177   2000
333.7—dc21   99-35703 CIP

ISBN 0 415 16829 5 (hbk)
ISBN 0 415 16830 9 (pbk)

Contents

Figures

Tables

Contributors

John Bachtler is Professor of European Policy Studies and Director of the European Policies Research Centre at the University of Strathclyde, Glasgow.

Chris Carter is Strategic Planning and Environmental Policy Manager for Worcestershire County Council.

Keith Clement is an independent environmental consultant and Senior Research Associate at the European Policies Research Centre, University of Strathclyde.

Jennifer Dixon is an Associate Professor in the School of Resource and Environmental Planning at Massey University, Palmerston North, New Zealand.

Joe Doak is a lecturer in the Department of Land Management and Development at the University of Reading.

Neil Ericksen is Director of the International Global Change Institute, University of Waikato, Hamilton, New Zealand.

David Gibbs is Professor of Human Geography at the University of Hull.

Andrew Gouldson is a lecturer in Environmental Policy at the London School of Economics.

Douglas Hart is Reader in the Department of Land Management and in the School of Planning Studies at The University of Reading.

Tony Jackson is a senior lecturer in the School of Town and Regional Planning at the University of Dundee.

Alexander Johnston was formerly Chief Planner at Leicestershire County Council and is now an adviser on planning and related matters.

James Longhurst is Professor and Associate Dean in the Faculty of Applied Sciences and Head of the Department of Environmental Health and Science in the University of the West of England, Bristol.

Angus Martin is a consultant with Pieda Ltd.

Jo Milling is a Planner in the Policy and Transport Section of Mendip District Council.

Dean Patton is a lecturer in Corporate Strategy at De Montfort University, Leicester.

Joe Ravetz is Director of the City-Region Centre at the Department of Planning at Manchester University.

Stephen Rees is a planner in the Environment Programme of the County Planning Department of Kent County Council.

Peter Roberts is Professor of European Strategic Planning at the University of Dundee.

Derek Taylor is a consultant on economic development and the environment with Global to Local Ltd.

Walter Wehrmeyer is a researcher at the Robens Institute at the University of Surrey.

Martin Whitfield was formerly a researcher at the University of Reading is now a consultant for an American I.T. solutions provider.

Elizabeth Wilson is Senior Lecturer in Environmental Planning at Oxford Brookes University.

Steve Winterflood is Head of Policy for Dudley Metropolitan Borough Council.

Ian Worthington is a Principal Lecturer in the Leicester Business School, De Montfort University, Leicester.

Preface

During the past decade considerable emphasis has been placed on the desirability of developing models and methods that encourage and support the integration of economic and environmental policy. Although the search for the perfect, fully integrated, model may prove to be an impossible dream, there have been a number of important advances in the theory and practice of integrated policy management.

However, despite this general pattern of advance, some sections of both the public and private sectors have ignored or resisted change. It is the intention of this book to demonstrate the benefits that flow from more closely integrating economic and environmental policy and, through this, to help to persuade the doubters and laggards of the merits of adopting an integrated approach.

Although a common and recognized body of theory is beginning to emerge, there is no single accepted theory that can be used to explain or guide the process of policy integration. Neither is there a common model of governance or administration that can set the standard of good practice. Instead, what emerges from the contributions to this book is a rich picture that demonstrates the wide range of current theory and practice.

In one sense the wide variety of theory and practice that exists is fortunate, because it means that various models are available that can be applied in a manner that fits individual circumstances. However, variety does run the risk of causing confusion and this may conceal or obscure good practice.

The contributions to this book demonstrate the existence of many common problems and offer some shared solutions. In selecting contributions we have attempted to balance depth against geographic variety, and we have attempted to encourage both academics and practitioners to share their views and experiences.

Although the overall coverage of the book is extensive, we do not claim that it represents any more than a snapshot of a rapidly evolving area of research and practice. Despite this limitation, we hope that it does represent a positive step towards the closer integration of economic development and environmental management.

Our thanks are due to a number of individuals who have assisted in the preparation of this book. First, to the contributors for their tolerance of our demands. Second, to our colleagues at the University of Dundee and the London

School of Economics for their assistance in the preparation of the manuscript; particular thanks go to Pam Gibb and to Cheryl Hardy-Theobald. Finally, our thanks go to the staff of Routledge for their forbearance and help.

Andrew Gouldson and Peter Roberts
London and Dundee, March 1999

Part I
Introduction

1 Integrating environment and economy

The evolution of theory, policy and practice

Andrew Gouldson and Peter Roberts

Introduction

In recent decades governments at all levels have introduced numerous environmental policies in an attempt to respond to the negative impacts of economic development. Many of these policies have been at least partially successful as it is likely that the quality of the environment would have deteriorated much more rapidly without such policy intervention. However, despite some important successes, it is generally accepted that the policies that have been introduced have not been able to address the negative impacts of economic development in a comprehensive way. This view is supported by Janicke (1997: 2) who argues that:

> success [in environmental policy] is more or less restricted to problems that can be handled (mainly by 'additive' technical standard solutions) without restricting markets or relevant societal routines. As regards the big environmental problems ... government failure and capacity overload can frequently be observed.

This quote introduces a number of the critical issues that will be addressed in the discussion that runs throughout this book. Particularly, it highlights the significance of certain forms of government failure and the need for institutional capacities to be developed that will allow policy makers to respond to the limits of existing approaches to policy intervention. However, while this book is centrally concerned with the ability of policy to influence environment–economy linkages, its primary focus is not on how environmental policy might restrict markets and societal routines. Instead, it considers how integrated approaches to policy might be developed and whether these integrated approaches might better enable economic and social activities to support rather than undermine the realization of environmental objectives.

The need for 'joined-up' thinking

In relation to the environment, government failure is typically seen to occur where policies are not introduced as a response to market failure, thereby allowing many environmental resources to be under-provided or over-exploited. While the classical interpretation of government failure emphasizes issues related to the level of environmental policy intervention, government failure also manifests itself in two other ways that are more related to the form of policy intervention. First, government failure occurs where the policies that are introduced as a response to market failure are environmentally ineffective or economically inefficient. In such instances it is commonly argued that new forms of policy intervention should be adopted that would allow some combination of higher environmental standards and lower economic costs of compliance to be realized. Second, government failure occurs where other policy areas, perhaps inadvertently, generate negative environmental impacts. This is readily apparent in the many instances where environmental objectives are not taken into account in the formulation, implementation or appraisal of policies that primarily seek to promote other objectives. This form of government failure is particularly significant as, despite the increasing priority that many governments claim to have awarded to environmental concerns, it is relatively unusual for environmental objectives to be fully integrated into policy areas with economic objectives.

For many environmentalists, the issue of integration is particularly significant as in many instances the peripherality of environmental objectives in the mainstream of economic policy making is seen to constrain environmental policy to a reactive role. The consequence of this is that environmental policies are commonly relied upon to respond to the negative impacts of policies in areas such as energy, agriculture, transport, trade and industry that are primarily designed to promote economic development. While it is acknowledged by many environmentalists that some environmental problems have been resolved or reduced by such a reactive approach, it is also argued that reactive forms of environmental policy have been and are perhaps increasingly unable to respond effectively or efficiently to many of the negative impacts of economic development. This view appears to be reflected in practice in the many instances where relative reductions in the impact of each unit of production or consumption have been more than offset by increases in the absolute level of economic activity. Furthermore, many environmentalists are concerned that once the easier policy options have been exploited, the remaining environmental problems are likely to be those of a more scientifically complex or politically intractable nature (see Christoff 1996).

These concerns about the performance of reactive environmental policies raise broader and possibly more fundamental questions about the nature of economic growth and development and its relationship with the environment. Commentators such as Giddens (1991) and Beck (1992) argue that modern industrial society is increasingly insecure about its future partly because it is questioning the faith that it has traditionally placed both in the market and the institutions of government to control the risks associated with economic development. Indeed, it is argued by Beck (1992) and others that the continued economic development upon which modern society appears to depend has either actually generated or has established the potential for new, more complex and increasingly pervasive environmental risks. It is also argued by some such as Braun

(1995) and Christoff (1996) that modern societies do not yet have and perhaps never will have the capacity to recognise and respond to many of the increasingly complex risks associated with economic development.

This discussion highlights the basis of calls for more proactive approaches to environmental protection that may enable policy makers to better anticipate and avoid the environmental risks associated with economic growth and development. In this respect it is clear that environmental objectives could be pursued more proactively if they were better integrated into the design and implementation of the various policies that seek to facilitate and promote economic growth and development.

While such a view may be supported by many environmentalists, many parties concerned with industrial growth and development are concerned about the economic implications of environmental protection. Clearly any discussion on strategies for integration must take economic as well as environmental interests into account. In this respect, theories of ecological modernization argue that economic and environmental concerns can be addressed simultaneously with the help of innovative forms of government intervention.

Ecological modernization

While ecological modernization is generally presented as a conceptual framework for understanding the potential of economy–environment integration, as a concept it has been interpreted and applied in three main ways. First, following authors such as Giddens (1991) and Beck (1992), it has been used to describe the ways in which modern society is responding to its increased awareness of, and anxiety about, the ecological risks associated with industrialism (see Mol 1995, Christoff 1996). Second, it has been used to describe and analyse emerging discourses in the debate on the environment and industrialism (see Hajer 1995). Third, it has been used in a prescriptive way as a concept to guide programmes of policy reform (see Gouldson and Murphy 1996; 1997). In this latter respect, rather than seeing environmental protection as a brake on growth, ecological modernization promotes the application of new forms of environmental policy as a positive influence on economic development. Similarly, rather than perceiving economic development to be the source of environmental decline, ecological modernization calls for the application of economic policies that harness the forces of entrepreneurship for environmental gain. Thus, in its prescriptive mode ecological modernization suggests that economic and environmental goals can be integrated within the framework of an advanced industrial economy through new and innovative forms of policy intervention (Gouldson and Murphy 1996). In this respect, ecological modernization calls for new forms of policy intervention that can induce changes that reduce many of the environmental impacts of economic development, for example by changing the spatial distribution, the sectoral balance, the technological composition or the resource intensity of socio-economic activity (see Gouldson and Murphy 1997).

In relation to the nature of economic development, the theory of ecological modernization looks for industrial sectors which combine higher levels of economic development with lower levels of environmental impact. In particular, it seeks to shift the emphasis of economic development away from energy- and resource-intensive industries towards service- and knowledge-intensive industries. It argues that changes in infrastructure and technology – such as through public-transport provision, land-use planning and the use of information technologies – can also make a major difference to environmental impact. At the micro-economic level, ecological modernization also assigns a central role to the invention, development and diffusion of new technologies and techniques. In particular, it seeks a shift away from reactive 'control' or 'clean-up' technologies towards the development and application of more anticipatory 'clean' technologies and techniques that tend to be associated with increased costs in the short term but that are widely held to be more environmentally effective and economically efficient in the medium to long term (see Gouldson and Murphy 1998).

Thus, through a combination of sectoral, infrastructural and technological change, in its prescriptive form ecological modernization proposes that the structure of economic development can be and should be reoriented to establish a more environmentally benign development path. Driven by new forms of policy intervention, this path would require the consumption of fewer resources and the generation of less waste, while also improving economic welfare and creating employment opportunities in labour-intensive rather than resource-intensive sectors.

Advocates of ecological modernization argue that such changes at both the macro-economic and micro-economic levels have the potential to make significant improvements in the environmental performance of industrial economies. Ecological modernization is thus presented as a means by which capitalism can accommodate the environmental challenge. Its advocates suggest that by accepting capitalism, but by seeking to reform it in technologically and politically feasible ways, ecological modernization at least provides some purchase for environmentalists in mainstream economic debate. Critics of ecological modernization, however, suggest that it offers only a partial and a temporary response to the causes of environmental decline: partial as it is silent on crucial questions of social change relating, for example, to social justice, the distribution of wealth and power and society–nature relations; temporary as it fails to address 'the crisis tendencies which are seen to be endemic in the [capitalist] accumulation process' (Gibbs 1996: 5).

Within the context of this debate, it is argued here that even if ecological modernization represents only a temporary point of convergence in an area of broader contestation between economic and environmental interests, at least this convergence establishes an opportunity for the potential of new forms of policy intervention to be explored. This is significant as the programme of policy reform prescribed by advocates of ecological modernization has yet to be adopted and consequently its potential has yet to be explored in practice. However, many of the avenues for policy development that are currently being explored by governments at various scales are at least compatible with the wider prescriptions of ecological modernization. Consequently, it is appropriate to examine the influence that new forms of policy can have on the relationship between environment and economy and the prospects for further change. In this respect, whilst acknowledging the broader conceptual criticisms of ecological modernization, this book seeks to examine the nature of those contemporary policy developments that have the potential to influence the relationship between environment and economy at the regional and local levels.

The local and regional context for integration

Although it is perfectly possible to design and develop policies at a transnational or national level that bring together environmental and economic concerns, such policies are frequently incapable of delivering specific 'products' or results. It is increasingly recognized that the detailed design and implementation of an integrated package of policies is more likely to be achieved at a regional or local level than at a higher level in the hierarchy of governance. In addition, it is also now accepted that the specific economic, social and environmental conditions that are experienced in an individual place may necessitate the development of a specific 'blend' of policy.

It is for these reasons, alongside the very practical requirement that expenditure should, whenever possible, allow for the achievement of more than one policy objective, that attention has increasingly been paid to the regional and local dimension in the search for ways of bringing together environmental and economic concerns. In many senses this is not a new discovery, rather it can be seen to represent the rediscovery of the merits of 'territorial integration' as a key organizing concept (Friedmann and Weaver 1979). Many of the earliest experiments in 'balanced development', an approach that was similar to what we nowadays refer to as sustainable development, were associated with programmes for local and regional development, often in 'natural' regions such as river basins. Geddes (1915) understood and explained the significance of the 'valley section' as a unifying force that can act as a container for an integrated package of policy, whilst in the modern era Cohen (1993) has argued that the region or city–region is the preferred level of government and governance for the achievement of sustainable development.

Above and beyond these basic concerns, there is a further, more practical matter that argues in favour of the adoption of a local and regional perspective on the integration of environmental and economic policy. This is, especially within the European Union, the increasing use of the region as a common building block for policy development and the distribution of financial support. Sustainable regional development is a matter that has come to the fore in recent years and the need to more closely integrate economic and environment matters is evident in recent revisions to the regulations which govern the operation of the Structural Funds. Equally, in individual nation states, economic development agencies are now frequently required to incorporate environmental objectives in their development programmes (Gibbs 1998). Sustainable regional planning and development can, therefore, be seen as a practical means of promoting the integration of economic and environmental concerns (Roberts 1998).

The focus of this volume

Given the apparent need for integration and the significance of the regional and local dimensions in pursuing integrative strategies, the contributions within this book set out to examine a range of factors relating to the context for integration, the motives and capacities for integration and the performance of those integrative initiatives that have been adopted to date. While international experiences with integration are addressed, the case studies within this book focus primarily on emerging experience with integration at the local and regional levels in the UK. Consequently, the cases reported in this book reflect a range of factors that are both spatially and temporally specific. However, particularly as the UK has been toward the forefront of the development and application of local environmental initiatives throughout the 1990s, it is hoped that a specific analysis of this nature can inform the wider debate on the integration of economic and environmental policies and plans at the local and regional levels.

Given the context for the integration of environmental and economic policies and plans outlined above, the contributions to this book consider a range of common themes.

First, they consider the motives for integration by assessing the political pressures for integration, the incentives and disincentives associated with integration, the distribution of the costs and benefits of integration both across social groups and over time and the influence of the different groups that support or oppose integration.

Second, they consider the local and regional capacities for integration by analysing the barriers to integration and the ways in which these barriers have been overcome, be it through the development of new institutions, structures and competencies, through the adoption of new approaches to the formulation or implementation of policies or plans or through the application of new procedures and practices.

Third, they consider the nature of the response by examining questions related to the form that strategies for integration have taken. Thus, the contributions consider whether calls for integration have led to the development of predominantly new initiatives that effectively integrate economic and environmental policies and plans or whether existing policies and plans have been revised to accommodate new concerns. Relatedly, the contributions consider whether calls for integration have led to the introduction of new structures and practices within local and regional authorities. In each instance, the contributions attempt to assess the reasons for the selection and adoption of particular approaches and the relative merits of alternative approaches.

Fourth, the contributions consider the performance of the adopted measures by looking at the social, economic and environmental consequences of integration. The contributions assess the influence of integration on the spatial and sectoral form of development and on the operations of the regional and local economy. Here issues related to the environmental efficacy and economic efficiency of particular policies and plans are examined.

Finally, the contributions consider the scope for future progress by examining whether the measures that have been adopted to date will continue to be applied, whether they can be refined and whether the experiences that have been gathered to date are transferable to other localities. The contributions also consider the need for further integration and the ways in which further integration might be better promoted or facilitated. The longer-term limits to integration are also discussed.

Aside from these key themes that will run throughout the analysis, the discussion is structured so that it begins with an examination of the wider (international) context for integration before moving on to consider the UK context for integration in more detail. It then assesses different approaches to and agendas for integration before considering issues related to the structures and systems needed to facilitate integration and indicators for the measurement of progress.

The structure of this volume

The wider context for integration

The perception of the relationship between environment and economy that underpins the contemporary policy framework is examined in the chapter by Keith Clement and John Bachtler which focuses on the approaches proposed and in some instances adopted at the supra-national level by the EU. This chapter examines the evolving nature of EU environmental policy and the influence that environmental objectives are beginning to have on the formulation, implementation and evaluation of non-environmental policies, particularly those designed to promote regional development. The discussion places contemporary EU policy developments into a historical context by assessing the gradual evolution of the EU's environmental policies over the last 25 years. It is argued that over this period the EU's environmental policies have gradually shifted from their original position where the primacy of economic development relegated environmental policies to a reactive role to their current position where environmental objectives are beginning to be integrated into the definition and application of those policies designed to promote both economic and social objectives.

In relation to regional development assistance, the chapter highlights the growing significance of new policy requirements that emphasize the need for the greater integration of environmental priorities into economic development strategies through more substantial involvement of environmental authorities, more effective screening of proposed regional plans and programmes and the identification and implementation of environmental criteria during the appraisal of specific projects at the local level. Whilst recognizing that at both the EU and the member-state levels very few countries have even assessed the environmental impacts of regional-development policies, the chapter suggests that there is a growing impetus for greater interaction between economic and environmental policies and that the EU and an increasing number of member states are stressing the significance of environmental factors within their regional-development policies. The chapter argues that progress has been made through initiatives designed to promote integration, to build more appropriate institutional structures and capacities, to develop new assessment methodologies and performance indicators and to ensure that implementation contributes to the realization of stated objectives. However, it also acknowledged that there is still a tendency towards the rhetoric of sustainable development at the strategic level whilst environmental actions continue to be compartmentalized within specific priorities or measures.

The evolution of environmental policy is also the focus of the chapter by Martin Whitfield and Douglas Hart which examines the dynamic tension between environmental protection and economic development and the complex interplay between federal, state and local policies in the USA. This chapter charts the evolution of environmental policies at every level in the USA, particularly since the late 1960s, and the success of these policies in at least partially reducing the level or the rate of environmental degradation. However, despite these practical successes, it also suggests that there is a growing recognition of the significance of the costs of compliance both by communities at the local level and by industries that must respond to the increasing demands of federal and state laws. Thus, the chapter emphasizes the need for new approaches that are more effective, more efficient, more participatory and more transparent so that the simultaneous and in some instances competing demands of local communities for economic development, social welfare and environmental protection can be balanced in a more accountable way.

Similar themes are addressed in the chapter by Jennifer Dixon and Neil Ericksen that considers the implications of the reform process that has dramatically influenced both the structures of government and the framework of policy that governs resource management in New Zealand. Once more, this chapter highlights the tensions that surround both the relationship between economic development and environmental protection and the balance between national and local policy and action. In this respect, it is argued that there is a degree of dissonance between the intentions of national policies designed to reduce the tensions between environment and economy by promoting an integrated approach and the realities of policy implementation at the local level. By way of an explanation, the chapter suggests that many local authorities in New Zealand are faced with a lack of coherent national policies, with multiple and in some instances competing objectives, with over-stretched resources and with under-developed capacities for implementation. Within this context, they highlight the possibility that standards will tend toward the lowest common denominator as local authorities compete to attract and retain economic activity. However, rather than welcoming local control, it is suggested that within industry there is commonly resistance to the variability of local approaches to environmental management and support for greater consistency and uniformity. Thus, particularly where there is overt tension between those promoting environmental protection and those who wish to secure the benefits of economic development, it is apparent that local initiatives can benefit from national strategies.

The UK context for integration

The discussion then moves on to focus particularly on the attempts to integrate environmental protection and economic development that many UK local authorities have adopted in recent years. Despite initiatives at the international, European and national levels, the chapter by Derek Taylor argues that in the UK, the progress of local authorities with integrating environmental with social and economic policies is patchy for a number of reasons. For example, he argues that as the concepts underpinning integration are new, difficult to internalize and not yet fully understood in practice, many local authorities are at the bottom of a steep learning curve. He also suggests that the impetus and incentives for integration are increasingly apparent with pressure emanating both from 'above' through programmes at the international, European and national levels and from 'below' where community concerns about the nature of development and the legitimacy of local government are being more clearly articulated. However, in response to these pressures he suggests that existing local government structures do not encourage integration in many instances and that there is commonly both an antipathy to new ideas and a resistance to change, particularly where they challenge existing structures, specialisms and interests. These barriers to change are compounded as, although integration is legally required in some areas, it remains largely a voluntary matter. Furthermore, he suggests that while there has been increasing pressure to integrate, calls for integration from national government have not been associated either with credible action or by consistent, coherent support from the centre. The pressure to integrate has therefore been accompanied by periods of friction and turbulence between central and local government and by increasing responsibilities and decreasing powers and resources at the local level. Nonetheless, Derek Taylor argues that many local authorities have taken the initiative and have begun to develop the necessary capacities and mechanisms to harmonize their environmental, economic and social objectives, policies and practices.

The capacity of local authorities in the UK to implement integrative policies is also the focus of the chapter by David Gibbs and James Longhurst. They argue that there are implicit assumptions within international and national policies about the capacity of local authorities to promote integration at the local level when in reality little is known about the structures and practices of local government that may act to facilitate or hinder such integrative activity. Through an empirical investigation of related UK local authority perceptions and activities, they conclude that while there is some evidence of a move towards integration, to date this has occurred at a fairly superficial level and is not yet a widespread phenomena. Indeed they suggest that the dominant trend is for economic development and environmental protection to remain separate areas of responsibility within UK local authorities. Significantly, they also conclude that where attempts have been made to integrate environment and economy, economic concerns appear to dominate so that integrative policies and initiatives actually form an 'add-on' to the main business of economic development. In part they link this to the lack of a clear vision about how a sustainable economy could be achieved, not only at the local level but also in the abstract guidance and vague recommendations issued by central government. Particularly, they suggest that integration requires methodologies for appraising and assessing proposals and plans and prioritizing developments that do not yet exist.

Many of the points raised by David Gibbs and James Longhurst are reinforced by the survey of the views and experiences of local-government environmental co-ordinators reported in the chapter by Stephen Rees and Walter Wehrmeyer. In common with many of the other contributions in this book, this survey found that there are some significant barriers preventing the development and implementation of integrative strategies at the local level. Thus, their survey suggested that sustainability initiatives have generally been adopted in UK local authorities against a background of limited top-level commitment, scarce resources, inappropriate organizational structures and restricted influence. Their survey also found that sustainability initiatives have been particularly influenced (and in some instances even captured) by certain professional groups within local authorities, notably those in environmental health, land-use planning and civil-engineering functions. They argue that this 'professionalization' of the sustainability agenda may serve to exclude the wider social and cultural aspects of sustainability and the groups that seek to promote social change. Critically, they also suggest that there is a crisis of confidence in many local authorities derived particularly from the perception that local authorities do not feel that they have the legitimacy or the credibility needed to exert a significant influence on the nature of development within their area. In response they argue that local authorities need to reassess their role in the communities they serve, to redefine their values and objectives and to develop new mechanisms and processes to translate the broad concepts of sustainability into practical initiatives for integration. In this respect they argue that local authorities need to foster partnership with the various actors in the local economy and community and to encourage the 'participation that is needed to improve the design and to enhance the accountability and credibility of their actions.

Integrating environment into development policies and plans

The various chapters on the context and the capacities for integration in UK local government stress the fact that local authorities have a limited influence on the broader nature of economic development and on its relationship with the environment. Thus, they suggest that a more coherent, consistent approach is needed that encourages a degree of synergy or resonance between the policies, plans and programmes brought forward at the supra-national, national and regional levels and the initiatives developed applied at the local level. The various contributions also argue that new processes and mechanisms need to be developed to assess the relationship between economic development and the environment and thereby to facilitate new ways of managing this relationship. The chapter by Elizabeth Wilson addresses both of these issues by focusing on the nature of Strategic Environmental Assessment (SEA) and the influence it might have on regional development policies, plans and programmes. In this chapter it is argued that there has been a growing recognition that policies brought forward at all levels can and do have adverse environmental consequences and that these consequences can be more effectively anticipated and avoided. However, whilst in the past attempts to anticipate and avoid the negative environmental impacts of economic development have tended to focus on the project level, it is suggested that the limits of the system of project-focused environmental-impact assessment (EIA) have become more readily apparent as experience with its application has accumulated.

Partly as a response to the shortcomings of EIA, SEA has been developed and is increasingly applied as a more formal, systematic and accountable way of evaluating the impacts of proposed policies, plans or programmes. Whilst it is acknowledged that there have been both political and technical problems with SEA, the chapter argues that the conditions for its wider and more successful application can be met as political support is secured, as the procedures for SEA evolve and as evidence of its technical feasibility accumulates. As was the case with EIA, experience with SEA is likely to accumulate over time, particularly if in some instances it becomes a legal obligation. Critically, where it has been applied, it is argued that it has at least begun to change cultural and institutional attitudes to economic development even if it has yet to exert a tangible influence on the nature of development at the local level.

The discussion then moves on to consider the wider lessons that can be drawn from the specific experiences of a number of initiatives concerned with the integration of economic development and environmental protection at the regional and local levels in the UK. The first case study from Chris Carter and Steve Winterflood considers a recent regional initiative that has attempted to facilitate the further integration of environmental concerns into regional-development plans and strategies in the West Midlands. As the chapter suggests that environmental concerns have become a more prominent issue in the preparation of regional plans, it considers the ways in which environmental concerns might be more effectively integrated into regional-planning guidance. The chapter then outlines the activities of a Regional Forum of Local Authorities that has sought to facilitate partnership between the various local authorities for the municipalities, counties and districts in the West Midlands. It has also sought to influence the nature of the regional-planning guidance issued by national government. In this respect, it suggests that the Forum has recognized and has argued that environmental rehabilitation, protection and enhancement should not only be seen as important goals in their own right but that environment should also be seen as an aim that should pervade and influence other regional objectives.

The chapter then examines the experience gathered in the West Midlands in its attempts to assess and reduce the environmental impacts of the policies that guide economic development within the region. Following a baseline review of the environment within the region, the chapter then describes the process through which the impacts both of current policies and of the various policy options were assessed. This assessment used a matrix-based approach that listed positive, negative and variable impacts whilst also acknowledging those areas where insufficient evidence was available to make a judgement. Based on this form of SEA, a range of policy proposals were developed. However, the authors suggest that many of the proposals put forward encountered political, financial and technical limits that precluded their adoption. These limits relate particularly to the need for extra resources to be found or diverted from existing uses to support new initiatives and to the need for a level playing field to reduce the ability of some regions and localities to ignore environmental objectives as they compete for inward investment. Despite these barriers to change, however, the chapter suggests that the process of assessment and the debate that has followed this process has led both to institutional changes as new strategies have been established and panels introduced and to cultural change as environmental concerns have been accepted as a more integral part of the regional-planning process and as the goals of this process have begun to be redefined.

The nature of the structure plan for Leicestershire, a county falling under the influence of the Regional Planning Guidance issued for the East Midlands, is then examined in a chapter by Alexander Johnston. This chapter reflects the current situation in the UK where the local response to regional-planning guidance depends upon the structure of government within the locality concerned. In areas with unitary authorities, the broader strategies of regional plans are operationalized within unitary development plans. In areas where there are two tiers of administration, county councils respond to regional plans by introducing structure plans that in turn guide the local plans introduced by district councils. At the most local level, local plans identify and set aside specific areas for particular forms of development. In such instances both structure plans and local plans must be consistent with the broader strategies introduced by the regional plan.

In the case of Leicestershire, the chapter suggests that the structure plan has introduced a strategic planning framework designed to accommodate economic growth, to improve employment opportunities and increase economic prosperity whilst giving increased emphasis to a better quality of life through the conservation and enhancement of the environment. The structure plan for Leicestershire focuses on a number of key areas, notably related to transport planning, housing and employment provision, environmental conservation and enhancement, derelict land reclamation, minerals and waste planning and management and a number of other priority areas such as the regeneration of city centres and areas affected by industrial decline. It is argued that the structure plan introduces a strategic focus for the various policies and plans brought forward by the different departments of the county council and by the other actors, particularly district councils, that influence the nature and spatial distribution of economic development within the county.

Within the structure plan for Leicestershire a variety of more detailed action plans for specific areas translate broader strategic objectives into more focused operational programmes. In formulating the structure plan, it is suggested that a debate can take place on the relative weight that should be given to economic and environmental concerns and on the ways in which conflict between the two can be reduced. By opening this debate to external scrutiny and wider participation it is argued that some consensus can be built so that the various social, environmental and economic actors in both the public and private sectors within the county can support or at least understand the approach to development advocated by the structure plan. Significantly, it is proposed that if economic development functions were to be divorced from planning and environmental protection functions, for example through the creation of a stand-alone economic development agency, such an integrated approach may not take place and the process of development may not be subject to sufficient local democratic control.

The following chapter by Jo Milling maintains the focus on the influence of development policies and plans on the relationship between environment and economy but this time examines experience at the local level. By focusing on local land-use plans, this chapter highlights once again the importance of coherence and consistency at the national, regional and local levels as national and regional policies are seen to both facilitate and constrain (and in some instances to contradict and undermine) initiatives at the local level. Rather than examining these issues further, however, this chapter focuses on how a local authority can comply with the requirements of regional plans and structure plans in a way that reflects the social, environmental and economic concerns of its constituents.

More particularly, the chapter discusses the development and application of a SEA methodology that fitted with the culture and capacity of the local authority. This methodology based itself initially on the preparation of a state-of-the-environment report for the area and subsequently on an iterative process that assessed and discussed the impacts of different policy alternatives to allow proposals to be revised and refined before they became more firmly established. At all stages in the process full and early public participation in plan preparation was encouraged to allow local communities to shape the plan. Such public participation and external scrutiny was facilitated particularly by the publication of matrices setting out the predicted impacts of proposed activities that were categorized as positive, negative, variable or uncertain and were linked to a commentary explaining how each assessment had been made. The authors suggest that this approach can be relatively quick, that it can introduce areas of environmental concern not previously considered, that it can be accessible to all and that it can make explicit the environmental choices being made.

Significantly, the assessment process discussed by Milling not only considered potential impacts but also sought to identify the capacity of different places to absorb those impacts. In this way the authority attempted to channel development spatially toward those areas with spare capacity rather than toward those areas where capacity had been or was expected to be exceeded. Where capacity had been or was expected to be exceeded, or where there was conflict between social, environmental and economic concerns, revisions were made to the proposed plan to avoid conflict where possible or to set out explicitly the reasons for the ultimate decision where conflict could not be avoided. Thus, the authors do not suggest that SEA is always able to promote or facilitate integration, only that it encourages a formal and systematic assessment and in some instances a revision of the various options. It also enables a more transparent and accountable approach to decision-making in instances where conflict between environment and economy persists.

Developing shared agendas for integration

Aside from the growing significance of initiatives designed to raise the status of environmental considerations in the development-planning process, in recent years Local Agenda 21 (LA21) has been an important influence on the structures, strategies, policies and plans of many local authorities. The chapter by Dean Patton and Ian Worthington reports the findings of a recent research project that examined the approaches to policy formulation and implementation in five particularly proactive local authorities. These authorities had developed integrated approaches to economic development and environmental protection as part of their LA21 initiatives. This research was based on the recognition that the notion of sustainable development that LA21 seeks to operationalize at the local level provides a basis for the integration of social, environmental and economic concerns.

Before examining the activities of the five proactive local authorities, a survey of the perceptions and activities of a wider range of local authorities was undertaken. Reinforcing the findings of the surveys referred to in previous chapters, this survey highlighted the significance of a number of factors that appear to be restricting the influence of LA21. These factors included the lack of tangible central-government support for LA21, the limited availability of financial resources, the lack of appropriate organizational structures and restricted cross-departmental support for integrative initiatives.

Within the five proactive local authorities that are considered in this chapter, however, these barriers to change had been overcome at least to some degree. In each of the case-study authorities, environmental concerns tended to have a high profile prior to the adoption of LA21 initiatives, largely because of individuals and/or groups who had championed the cause for some time. Where the importance of environmental concerns was already acknowledged, the authors suggest that LA21 had reinforced rather than challenged the pre-existing culture and commitment of the authority. In a number of cases commitment had also been increased as a consequence of participation in schemes such as the Environment City campaign or the National Sustainability Indicators Project.

Significantly, each of the five authorities was able to translate its commitment into action through structures that enabled co-operation both internally between departments and sectors and externally between the authority and other authorities and agencies and with groups representing social, environmental or economic interests. Thus, the authors suggest that success is more likely where well-established fora of local interests exist and where a network of formal and informal relationships enables broad participation by the different vested interests so that all actors and agencies can work together to prepare, promote and act on a shared LA21 agenda. However, the authors also note that support for LA21 initiatives is often conditional, particularly in the private sector where they suggest there is considerable reluctance on the part of many businesses to lend support to an initiative that is seen as relatively unimportant and of little immediate benefit. In industry, as well as in local government, an emphasis on short-term private costs rather than medium- to long-term social, environmental and, in some instances, economic benefits commonly precludes strategic action.

Issues related to participation, particularly from business interests, are explored further in the chapter by Joe Doak and Angus Martin. This chapter suggests that the simultaneous social, environmental and economic demands of sustainable development need to be managed through a participatory decision-making process that builds consensus both on the nature of the problems and the selection and application of responses. Critically, they suggest that consensus building needs to identify areas of mutual gain and attempt to build 'win–win' outcomes from the decision-making process. It is suggested that if such 'win–win' outcomes can be generated, the various interest groups whose support is needed can be drawn into the decision-making network to participate in the development and implementation of a shared agenda for sustainability. The authors suggest that such a process of consensus building can allow a common understanding of the concept of sustainability to develop where it may otherwise be contested, that it can aid implementation by promoting a better sense of ownership than traditional 'top-down' impositions and that it can mobilize wider support and allow a long-term strategic focus rather than the short-termism that may otherwise pervade. Significantly, they recognize that consensus building can also lead to temporary, compromised policies that reflect the interests of the more influential groups and that lack an institutional framework for implementation.

By focusing on the experience of the Lancashire Environmental Forum as it has attempted to encourage participation and build consensus to underpin the Lancashire Environmental Action Programme, the authors are able to analyse many of these critical issues. While many of the points that are raised by the authors through their examination of experience in Lancashire resonate with those put forward in other chapters, they focus particularly on the issue of industry's participation (or non-participation) in the local decision-making process on sustainability. They suggest that businesses are a vital instrument for implementation and that without their involvement the credibility of the process can be questioned. Thus, it may be that by collectively withholding their support businesses have the power to restrict and possibly even to undermine the influence of local initiatives for sustainability. In the case of Lancashire, the chapter suggests that while some representative organisations such as chambers of commerce have participated, industry in general and individual businesses in particular have been weakly represented. In part it is argued that this is because businesses perceive the County Council, which is the lead body in the process, as having a strong environmental and anti-industry stance. It appears therefore that the 'ownership' of initiatives is an issue. However, while the authors suggest that the lack of independent mediators to facilitate the process of consensus building has had a detrimental impact, they also recognize that without the strong political leadership of the County Council the process would not have been initiated or sustained.

The authors also suggest that business interests are able to exert more influence on the environmental policies that affect them by participating in debates at the national or regional rather than the local level. Thus, it is argued that businesses would be more willing to participate in local initiatives if more power and resources were devolved toward the local level. Indeed, they suggest that the increasing emphasis on the plan-led system of development, coupled with the weight that is now being placed on locally generated strategies by national and European funding programmes and the recent proposals for Regional Development Agencies, indicate that such a transfer of power and resources may indeed be taking place at least to some extent. Rather than transferring power to the local level, however, they suggest that decision-making at the regional level may offer the best opportunity to marry the globalism of economic and environmental processes with the localism of 'meaningful' participation by ordinary people.

Structures, systems and indicators for integration

The subsequent chapter by Tony Jackson changes the direction of the discussion a little to focus on the structures, systems and indicators needed to operationalize strategic objectives. This chapter describes the attempts undertaken by Fife Council to shift from a position where environmental concerns were addressed in parallel with its other more traditional activities to one where the broader concept of sustainability was incorporated into its strategic objectives and integrated into all of its departmental functions and sectoral activities. More particularly, the discussion focuses on the structures and systems introduced to facilitate integration and the performance measures and indicators developed to evaluate and communicate progress. However, once again the chapter stresses the importance of establishing a common understanding of the meaning of sustainability and therefore the need for consultation, participation and partner-ship when establishing strategic objectives and pursuing implementation.

In terms of the internal structures and systems of local authorities, the chapter suggests that a systems approach to management is better suited to the holistic demands of sustainability than the hierarchical structures common to most local authorities. It argues that these hierarchical structures tend to introduce and reinforce sectoral specialisms for individuals, departments and organizations. It also argues that these sectoral specialisms are typically associated with inappropri-ate performance measures that may give a false impression of progress if compartmentalized measures fail to recognize that progress in one area has been achieved at the expense of another. It is also suggested that as the performance of systems commonly depends upon the interaction between the various elements as well as the nature of the elements themselves, systems commonly generate unexpected and sometimes counter-intuitive outcomes. Therefore, it is argued that a systems-based approach to management is more appropriate as it can better promote vertical and horizontal integration in organizations and their policies, plans and programmes based on a recognition of the mutual dependence of the various elements and the need for co-operation between these elements.

Given this argument in support of a systems-based approach to management, the chapter discusses the contribution of management-systems standards such as that introduced in the EU by the local authority Eco-Management and Audit Scheme. Here, it is argued that experience gathered through Fife's participation as a pilot authority for EMAS supports the previous points as EMAS has helped the authority to pursue integration more effectively by helping to break down the functional boundaries that in the past have engendered division between its various activities.

Whilst stressing the importance of management systems, the chapter also highlights the significance of appropriate performance indicators. In this respect it is acknowledged that systems-based approaches such as EMAS are merely management tools and that their influence on performance depends upon the manner of their application. Based on experience gathered through Fife's participation in a pilot project on sustainability indicators, the chapter suggests that new indicators are needed which accurately reflect progress toward or away from sustainability. Once more it is argued that once a common view of sustainability has been established, participatory approaches to the development and use of indicators are needed to secure their legitimacy and credibility. It is also acknowledged that the various interest groups can also be usefully involved in performance measurement. Thus, it is argued that it is possible to establish a clear strategic vision that can be operationalized through an effective management system and that progress can be measured and feedback provided using appropriate indicators that reflect the multi-faceted, socially constructed and politically contestable nature of sustainability.

Conclusions

Whilst the desire to promote economic development remains the dominant policy goal for governments at all levels, there appears to be widespread public concern about the unanticipated consequences of continued industrial development. It can be argued that these concerns are directly related to the perceived inability of government and industry to anticipate and avoid new threats to the quality of life and the viability of ecological systems.

Calls for the fuller integration of environmental objectives into economic policies stem directly from the real or perceived limits of reactive approaches to environmental protection. Many people argue that such integration is necessary if economic policies are to reinforce rather than undermine environmental objectives. This book is based on the recognition that such integration can be effectively pursued at the regional and local levels where the need for 'joined-up' thinking is readily apparent. To date however, integrated approaches to economic and environmental policy-making remain very much the exception rather than the norm, even at the regional and local levels.

In response to these issues, this book examines the changing context for integration at a number of levels and locations and the experience with integration at the regional and local levels in the UK. To some extent the discussion that it presents is merely a snapshot of a rapidly evolving body of experience that is accumulating through the application of policies and practices that are necessarily diverse as they seek to take into account regional and local conditions and contingencies. However, it is hoped that the experiences reported in this book will also be of wider relevance as other regional and local authorities consider the need for integration and the nature of the various strategies for integration that might be adopted. It should also be acknowledged that the primary emphasis of this book, namely the debate on the integration of environmental objectives into economic policies, is closely related to a very similar discussion on how economic development might more effectively serve social objectives. The focus of this book on environment–economy relations is not in any way intended to downplay the significance of social objectives and the need for balanced and sustainable forms of economic development that further social and environmental as well as economic interests.

References

Beck, U. (1992) *Risk Society: Towards a New Modernity* (English translation), London: Sage.

Braun, E. (1995) *Futile Progress: Technology's Empty Promise*, London: Earthscan.

Christoff, P. (1996) 'Ecological modernisation, ecological modernities', *Environmental Politics* 5(3), 476–500.

Cohen, M. (1993) 'Megacities and the environment', *Finance and Development* 30(2), 44–7.

Friedmann, J. and Weaver, C. (1979) *Territory and Function*, London: Edward Arnold.

Geddes, P. (1915) *Cities in Evolution*, London: Williams and Norgate.

Giddens, A. (1991) *The Consequences of Modernity*, Cambridge: Polity Press.

Gibbs, D. (1996) 'Integrating sustainable development and economic restructuring: A role for regulation theory?', *Geoforum* 27(1), 1–10.

Gibbs, D. (1998) 'Regional development agencies and sustainable development', *Regional Studies* 32(4), 365–8.

Gouldson, A. and Murphy, J. (1996) 'Ecological modernisation and the European Union', *Geoforum* 27(1), 11–21.

Gouldson, A. and Murphy, J. (1997) 'Ecological modernisation: Restructuring industrial economies' in M. Jacobs (ed.) *Greening the Millennium? The New Politics of the Environment,* Oxford: Blackwell.

Gouldson, A. and Murphy, J. (1998) *Regulatory Realities: The Implementation and Impact of Industrial Environmental Regulation,* London: Earthscan.

Hajer, M. (1995) *The Politics of Environmental Discourse: Ecological Modernisation and the Policy Process,* Oxford: Clarendon Press.

Janicke, M. (1997) 'The political system's capacity for environmental policy', in M. Janicke and H. Weidner (eds) *National Environmental Policy: A Comparative Study of Capacity Building,* Berlin: Springer-Verlag.

Mol, A. (1995) *The Refinement of Production: Ecological Modernisation Theory and the Chemical Industry,* The Hague: Koninklijke Bibliotheek.

Roberts, P. (1998) 'Ecological modernisation: A model for future urban and regional planning and development', in P. Kivell, P. Roberts, and G. Walker (eds) *Environment, Planning and Land Use,* Aldershot: Ashgate.

Part II
The wider context for integration

2 European Union perspectives on the integration of environmental protection and economic development

Keith Clement and John Bachtler

Introduction

The past two decades have seen a revolution in approaches to regional and local economic development across Western Europe. Whereas the locus of responsibility for promoting structural change and development has moved from central government to sub-national tiers of government, arising from both political and economic pressures, new forms of territorial management have also arisen, partly through the top-down devolution and deconcentration of national government powers and partly from 'bottom-up regionalism', as local and regional communities asserted their economic, political or cultural identity and autonomy. As a consequence, the nature of regional institutions and responsibilities now varies substantially across Western Europe, encompassing state governments in Germany or Austria, autonomous regions and communities in Belgium and Spain, elected regional councils in France and Italy, and decentralized regional offices of central government in Finland and the UK.

The new models of territorial management evident across Western Europe embody several distinctive features. They emphasize 'inclusivity', taking account of the needs of economically or socially disadvantaged, marginalized and excluded groups such as women, the long-term or young unemployed, ethnic minorities and the disabled. They promote co-ordination and co-operation, both vertically between tiers of government and horizontally among regional and local actors; and they involve more strategic, integrated thinking in the planning and implementation of regional development. The nature of regional and local development has also become more complex, as the former reliance on various forms of planning and relatively simple instruments such as investment incentives has been superseded by a mix of measures promoting business development, physical and economic infrastructure, research and innovation, human resources, environmental improvement and community development. Lastly, regional and local action now proclaims a concern with sustainable development, integrating an environmental dimension into policies and programmes.

This chapter is concerned with the growing environmental commitment within the European Union (EU), focusing in particular on regional-development programmes co-financed by EU structural and cohesion policies. Although these structural policies form only one aspect of regional and local development, and in some countries provide relatively small amounts of expenditure in comparison with member-state policies, the procedures involved in programming EU expenditure provide a useful laboratory for analysing strategic approaches to economic development in different countries. Though variable in quality, the programming documents (Community Support Frameworks, Single Programming Documents) and evaluation studies required by the Structural Funds regulations provide insights into the nature of regional and local-development problems across the EU and the strategies and measures being adopted to address them. Given that the European Commission has promoted environmental awareness and concern over the past decade, these data sources provide some comparative empirical evidence of how different countries and regions approach environmental integration (in theory and practice) within development strategies.

The European Union and the environment

Environmental integration has been on the EU agenda for many years. It was introduced initially through the Environmental Action Programmes (EAPs), which have progressively promoted environmental policy as a positive contributor towards the solution of economic issues such as unemployment, economic growth and job creation (Hildebrand 1993). Emerging in the Third EAP (1982–6), it was stated that 'the EU should seek to integrate concern for the environment ... and this should result in greater awareness of the environmental dimension' (CEC 1983). The Fourth EAP (1987–92) was more forceful, indicating that as key factors in economic decision-making, environmental-protection policy and strict environmental-protection standards were no longer optional extras. It further stated that integration with other Community policies should be a central part of the Commission's efforts; this would be applied first in EU policies, then in member-state policies, then more generally so that all social and economic developments in public and private sectors would exhibit this integration (CEC 1987).

By the advent of the Fifth EAP, the emphasis on environmental integration in sectoral and economic policies was paramount (Johnson and Corcelle 1995). Entitled 'Towards Sustainability', the Fifth Programme relates to the period 1993–2000. In this document, the European Commission states that sustainable development entails 'preserving the overall balance and value of the natural capital stock, redefinition of short, medium and long-term cost/benefit evaluation criteria and instruments to reflect the real socio-economic effects and values of consumption and conservation, and the equitable distribution and use of resources between nations and regions over the world' (CEC 1993a). In contrast to the previous programmes, 'Towards Sustainability' emphasizes proactive measures; that is, an anticipatory policy tackling problems before they occur. In this connection the 'precautionary principle' is introduced as a cornerstone alongside the 'polluter pays principle' (formulated as a EC Recommendation as early as 1975). Moreover, it seeks to raise public awareness to change attitudes and behaviour towards environmental issues, broaden the range of instruments and integrate environmental considerations into other policy areas, especially the five sectors of manufacturing, agriculture, energy, tourism and transport.

In 1996, the European Commission published both a progress report on the Fifth EAP (CEC 1996a), containing a comprehensive appraisal of the developments since 1992, and an action plan that sets out the priorities for the years ahead. The action plan maintains the focus on the five target sectors (industry, energy, transport, agriculture and tourism) considered to deserve special attention because of their disproportionate contribution to current problems. Of these sectors, manufacturing industry received a favourable review in terms of its progress towards sustainable development, and it was given future goals in the form of improved resource management and the provision of better information to facilitate more environmentally friendly consumer choice. Relatedly, the Commission plans to establish stricter standards for products and production methods; and it is hoped that the enforcement of best available technologies will enable European companies to gain a competitive advantage in world markets. For the energy sector, the main objectives were given as improved energy efficiency and the use of fewer carbon-intensive energy sources, attempting to reduce the consumption of non-renewable resources as well as emissions of gases that contribute to a range of environmental problems. However, the EU-wide approach favoured by the Commission – the introduction of a combined carbon/energy tax – was rejected by the Council of Ministers.

With regard to the transport sector, the review of the Fifth EAP acknowledged that little, if any, progress had been made towards sustainable development considerations. Furthermore, medium-term forecasts show rapidly rising traffic volumes, so that environmental problems caused by congestion-associated pollution and vehicle emissions will become increasingly urgent; the Union's carbon dioxide reduction targets are also endangered by these transport developments. Accordingly, the Commission now seeks to balance public and private transport, and it suggests improvements to the European railway network, inter-modal co-operation and stricter measures such as road pricing or increases in fuel prices.

In agriculture, the key elements of the Commission's strategy to achieve harmonious coexistence of agriculture and the environment comprise stricter controls over the use of pesticides, herbicides and fertilisers as well as financial incentives to encourage more environmentally friendly farming practices. Nevertheless, in the near future the most significant developments in this sector will still predominantly reflect the evolution of the Common Agricultural Policy (CAP). Lastly, in tourism, the diversification of tourism activities and the promotion of environmentally friendly ('sustainable') tourism are seen as remedies for developments often associated with severe negative environmental impacts, especially in sensitive areas such as the Alps and other favourite destinations of mass tourism, mainly the Mediterranean regions.

When considering future European environmental policy, the Commission intends to strive for a deeper integration of environmental considerations into all public and private consumption and production decisions. Especially with regard to European regional policy, an earlier communication recognized the necessity to ensure that 'all Community funding operations and, in particular, those involving the Structural Funds, will be as sensitive as possible to environmental considerations' (CEC 1992a). Possible means of achieving this would include methods such as greater integration of environmental priorities into economic development strategies through more substantial involvement of environmental authorities, more effective environmental screening of regional plans and programmes, and the identification and implementation of environmental criteria during the appraisal of specific projects.

Legislative factors leading up to the Fifth EAP included the Single European Act 1987, amending the Treaty of Rome with the introduction of Article 130R(2), that environmental protection requirements should be a component of the Community's other policies. Following this, all Directorates General of the European Commission were under a legal obligation to consider environmental matters. The Maastricht Treaty subsequently strengthened Article 130R to state that environmental-protection requirements must be integrated into the definition and implementation of other Community policies.

The most recent EU economic initiative to consider environmental factors was the White Paper on Growth, Competitiveness and Employment (CEC 1993a). The principal focus was the economic and sustainable development of the EU member states, with a mission to create 15 million jobs by the end of the century and to resist international competition. With regard to environment, the orientation was restricted to how jobs could be created through environmental protection, and in general the document addresses more the interdependence of economies rather than the interdependence of economy and environment.

However, the White Paper also introduced the prospect of a new development model. Acknowledging that the conventional measurement of Gross Domestic Product (GDP) is losing its relevance for future policy design, the new development model seeks to reverse the current negative relationship between growth and pollution. It would directly address the structural links between environment and employment – such as the inefficient or excessive use of natural resources in the Community – that represent hidden welfare losses. Clean technology would be seen as a key to decouple the bad economic–ecological relationship, promoting longer-life products, increased efficiency, recycling and reduced production of waste; and in this context, new technology would form a primary and important input rather than focusing on outputs such as cleaning up waste. The anticipated secondary benefits include improved competitiveness and the demonstration of how sustainable development could be put into practice.

The role of environment in member-state regional-development policies

In terms of environmental protection, regional-development policies can have positive and negative impacts. Positive impacts or likely improvements to the quality of the environment might include outcomes such as promoting more effective use of resources, reducing congestion in overcrowded areas, enhancing the environment in declining urban and industrial areas, providing peripheral regions with pollution-treatment facilities, assisting firms to meet environmental standards and supporting activities that preserve rural areas. The opposite perspective relates to identifiable negative impacts of regional development, with less desirable outcomes such as directly financing activities that produce environmental damage, increasing the geographical spread of activities leading to higher energy consumption, multiplying transport links between peripheral and central locations, and contributing to reduced efficiency in the operation of central urban regions.

Within the member states of the European Union, very few countries have made assessments of the positive or negative impacts of regional policy on the environment, and the practice of integrating regional and environmental policy is still a very new phenomenon. Although survey work by the OECD has indicated that the United Kingdom and France have carried out assessments, these studies were incomplete and primarily qualitative rather than quantitative (Larrue 1994); and a study at regional level has revealed considerable inconsistency between approaches to environmental appraisal in English regions, even between authorities receiving identical guidance (Seamark 1996). In practice, very few countries have announced formal initiatives to move towards environmental integration, although Germany and the Netherlands have none-theless developed forms of regional environmental policy (Clement 1994). In most cases, the impetus for policy interaction between the economic and environmental sectors has resulted from initiatives by environmental-policy agents, such as the adoption of new environmental plans, new environmental legislation or the creation or strengthening of central administrative structures relating to the environment.

Nevertheless, in some instances, maintaining and enhancing environmental quality has featured as a fundamental component of regional policy, especially in programmes for declining urban and industrial areas. The improvement of environmental conditions to facilitate the attraction (and retention) of companies and inward investment has resulted in its inclusion within development programmes in a range of countries (OECD 1990), and this approach is expected to continue in future, particularly in the older industrialized regions of countries in transition and the new Länder of eastern Germany. However, this status of 'prerequisite for development' generally takes the form of a reactive approach, responding to the environmental damage already caused by financial support given for development in the past. As the consideration of environment is consequent upon earlier negative impact, these clean-up measures are realized through elements of programmes perceived as distinct from the main thrust of the initiative. Rather than a forward-looking integrative approach, which would elevate environmental budgetary allocations and objectives to a status similar to the economic measures, the environmental improvement focus is subsidiary to the conventional development orientation.

In some European countries, mechanisms have been designed to ensure that projects subsidized through regional policy meet both regional and environmental criteria. Examples include preferential treatment for projects that improve the environment and the use of environmental assessment at project and strategic levels. Active promotion of projects that improve the quality of the environment takes place in Greece – where tourist-investment projects are eligible for assistance only in areas where environmental capacity has been exceeded – and more widely in Austria, France, Greece, Italy and Portugal, each of which gives priority to projects encompassing pollution control or energy savings. Interestingly, in Switzerland, bordering on the EU territory, priority for assistance in the mountain areas is given to waste-disposal facilities and small power plants using renewable-energy sources and only latterly to transport facilities or artificial-snow plants. With regard to technology improvement, assistance may be provided for projects specifically related to the development of clean technologies that generate energy savings, such as happens in Austria and Belgium, and technological innovation centres established in subsidized regions of Lower Austria now specialize in clean-technology development (Larrue 1994).

The technique of environmental-impact assessment (EIA) represents one practical method of attempting to integrate environmental factors into regional policy. However, EIA is generally found to be a limited instrument in which environmental effects are not fully evaluated, and in which greatest attention is given to short-term effects. Long-term and indirect effects need to be assessed through broader techniques such as strategic environmental assessment (SEA), a common methodology for which is still under development (Therivel *et al.* 1992). SEA considers the consequences of a range of actions early-on in the planning process; it enables decision-makers to choose the most appropriate action on environmental as well as socio-economic grounds; and it seeks to minimize any remaining impacts. In practice, SEAs may take several forms: for example, sectoral, focusing on transport plans or programmes or energy policies; spatial, assessing economic programmes or development plans set at national, regional or local levels; or indirect, environmentally evaluating initiatives for themes such as research and development (R&D) programmes or plans for privatization of public-sector industries.

Within the European Union, attempts have been made to develop a form of SEA which would apply to Structural Funds programmes, amongst other initiatives. In 1991, the European Commission's Directorate General XI prepared an internal proposal for a draft directive, but a number of member states successfully opposed its realization (Wilson 1993). Currently, a new draft directive on SEA with a focus on plans and programmes is being reviewed by the member states (CEC 1996b).

EU regional development strategies and the environment

At the level of EU programmes, the reform of the Structural Funds in 1988 heralded a new approach to European regional policy. Building on the pioneering 'integrated development operations' launched in regions such as Strathclyde, Groningen and Naples, it required a more strategic approach to regional development based on multi-annual programmes combining the use of EU, national and local-policy expenditure. The new Structural Funds regulations also demanded a partnership approach to regional development between the European Commission, national governments and regional/local authorities. Member-state authorities were tasked with preparing a regional development plan for each eligible area designated under Objective 1 (development of lagging regions), Objective 2 (conversion of old-industrial regions) and Objective 5b (development of rural areas). These plans provided the basis for negotiations between the member state and the Commission, resulting in an agreed Community Support Framework for each region, specifying expenditure priorities and governing the subsequent allocation of aid through so-called operational programmes.

In preparing regional development plans, member states were given several environmental obligations under the Structural Fund regulations: to undertake a regional-environmental appraisal, to evaluate the expected environmental impact of the plan and to comply with EC environmental directives and regulations. Environmental authorities were to be involved in both the preparation and the implementation of the programme. In practice, environmental considerations were marginalized in these early regional development programmes, attributed to the challenging task (for member state authorities and Commission services) of interpreting the new Structural Fund regulations, the new concept and terminology of programming, and the tensions created by the bureaucratic procedures involved in preparing and negotiating programmes. Delays were inevitable, and – in the worst cases – programmes intended to run from 1989 for 3–5 years were only agreed in late-1991 (Bachtler and Michie 1994).

Against this background, the programmes were limited in scope, focusing heavily on business aid and infrastructure support. Environmental protection was subordinate to economic priorities such as job creation, and environmental measures typically took the form of land regeneration and landscaping (Woodford 1991). In 1992, the European Court of Auditors reported little sign of environmental conformity within the Structural Funds, and concern was expressed about the lack of strategic thinking on the role of sustainable development and the need for more fundamental integration of environmental considerations within regional development plans and programmes (CEC 1992b).

Focusing on programmes for older industrial regions (Objective 2), Structural Funds administrators are now emphasizing a more positive and proactive approach to environmental factors. Two main themes are to be pursued more vigorously in new Objective 2 programmes. The first relates to the traditional approach of improving the physical environment to increase the attractiveness of the region for business development; the second is more forward-looking, seeking future competitive advantage by exploiting eco-products, environmental services, environmental technologies, energy-saving measures and improved production processes (CEC 1993b). To illustrate progress towards these objectives, the following sections summarize the results from a recent environmental appraisal of the management and implementation of single-programming documents (SPDs) for twelve Objective 2 programmes in Austria, Denmark, Finland, France, Germany, Sweden and the United Kingdom (Clement and Fitzgerald 1997). The programmes are considered from the three orientations of integration, management and implementation.

Integration

Amongst programme partnerships, there is a growing awareness and understanding of environmental issues. A form of 'cultural change' is underway with respect to the role of environmental objectives, priorities and measures, although to varying degrees. In accordance with theories of ecological modernization (Gouldson and Murphy 1996), the fundamental and essential links between environmental issues and industrial development are being acknowledged explicitly. This perception has a direct practical influence on programme design, but environmental objectives can still subsequently be diminished in importance by administering authorities in favour of economic-development priorities.

In some cases, the prospect of integrating environmental issues into economic-development strategies represents a major step for participating authorities. In the UK, the West of Scotland partnership has succeeded in making the transformation from regarding environmental features as restricted to obvious physical measures – such as tree planting, 'sites and premises' and derelict-land reclamation – to perceiving them in more broad-based terms encompassing energy efficiency, environmentally friendly materials and transport modes. Similarly in Finland, previous experience with programme implementation has resulted in environmental factors receiving much greater attention in the new programme, reflecting a considerable increase in the use of environment-related expertise.

Factors determining environmental integration relate partly to strict regulatory frameworks, legal culture and pressure from environmental authorities. In Austria, Denmark and Germany, national environmental regulations and authorities are considered very strong, and this reflects a broad general awareness of environmental issues. Procedural safeguards are also considered very good, and – in the Danish and Austrian contexts – this has reduced the amount of environmental content considered necessary for inclusion within programmes.

Elsewhere, environmental organizations have played key roles. In France, important influences came from the European Commission, which was involved especially in framing environmental policies in the Rhône-Alpes programme, and from DATAR (Délégation à l'Aménagement du Territoire) through information dissemination at the beginning of each round of plan preparation, providing summaries of French SPDs and highlighting examples of best practice under a variety of headings, including environment. In formulating the North-Rhine Westphalia programme in Germany, the entire SPD was discussed with the State

Environment Ministry, following which the Ministry assumed responsibility for implementation of certain environmental components of the programme and overseeing procedures down to project level.

For industrial South Wales, the Welsh Environment Agency was instrumental in reviewing the SPD using the methodology of strategic environmental assessment to indicate that the programme had the potential to produce both beneficial and detrimental impacts. In this instance, the approach significantly modified and improved upon the standard methodology which the Welsh Office partnership provided for regional offices. In comparison, Strathclyde European Partnership engaged in a productive debate with Friends of the Earth and the Confederation of Scottish Local Authorities on the question of economic development and sustainability and how it could affect the west of Scotland.

Management

One of the more challenging aspects of securing environmental integration concerns the institutional arrangements utilized to manage the programmes environmentally, and the implications that these arrangements might have. In practice, environmental management within Objective 2 programmes generally reflects existing structures, national regulatory systems or administrative schemes.

Predictable interactions between economic development and the environment are evident in the Nordic countries, reflecting the advanced legal and institutional frameworks within which the programme partnerships operate. For the Fyrstad programme in Sweden, the County Administrative Board (CAB) – acting as the competent environmental authority – followed its involvement in SPD preparation with representation on the Programme Management Committee. As a matter of procedure, every project is sent to the environmental departments in the CAB for approval. This has not resulted in project applications being rejected, essentially because of the existing regulatory high standards. A similar situation exists in Denmark, with all projects that might potentially affect the environment being sent to the (county) Department of the Environment. Again, the emphasis is on ensuring that the proposed development does not contravene existing environmental regulations, and consequently this stage might be regarded as a necessary formality rather than be expected to reveal new insights.

In Finland, the environmental authorities participated in the drafting of the regional programme and in evaluating environmental impacts, and the national Ministry of the Environment has responsibility for overseeing aspects of environmental management. Independent working groups operate at regional level (comprising regional councils and environment district offices), where programme environmental management is carried out by the same teams that prepared the programmes.

An alternative approach to environmental management is to devise specialist complementary systems for steering economic development towards environmentally advanced practices, although such a structure has both advantages and disadvantages. In the example of Austria, SPDs are administered by programme managers, but thereafter monies are distributed through funding agencies and departments using existing schemes. This means that programme environmental management will be carried out by a combination of departments that administer relevant schemes. Whereas all projects must necessarily comply with environmental regulations, current specialist environmental schemes include support for industrial waste/sewage measures (Styrian State government), assistance for environmental-protection measures (Lower Austrian government), and two parallel schemes from the Ministry of Environment which are administered by the Österreichische Kommunal Kredit (OKK), an Austrian bank. In practice, these schemes are differentiated by their restricted eligibility for environmental investments that exceed minimum legal requirements. Although forward-looking in design, this form of scheme-dependence has in effect slowed programme implementation, as the European Commission required lengthy periods to approve the Styrian and OKK environmental schemes, in effect delaying their realization until approximately 2 years after the programme was launched.

Implementation

One of the drawbacks of current environmental practice in Objective 2 regions is that strengths in environmental integration and management are generally not carried through to the implementation phase. Project selection and programme monitoring are the two most significant issues.

Project-selection criteria mostly reflect reactive (or even passive) approaches to programme environmental management. An example from Denmark is the North Jutland programme, which restricts environmental criteria to general environmental requirements, and it includes no proactive promotion of projects with a positive environmental impact. The involvement of the Department of the Environment is only to ensure the legality of projects, not to consider setting priorities. However, within the programme management, some projects with a special environmental focus have been given a higher priority.

In France, there is a standard national approach to carrying out environmental-impact assessments, so there is no separate discussion of the approach to impacts in EU Structural Funds programmes. In the Aquitaine programme, all project applicants must indicate the likely environmental impact of projects, but the main question asked about proposals is whether or not they respect the rules with regard to environmental standards. It is considered to be beyond the remit of the organization to insist on the environment being accommodated in a more holistic way. For the Rhône-Alpes region, by contrast, measures with a positive environmental impact are set out in the programme's environmental profile, but negative environmental impacts are not identified.

The exception within this category is the West of Scotland programme, where recent revisions have incorporated criteria for environmental management within the scoring system devised for project selection. Applicants who fail to complete or to return this section of the form automatically lose points from their potential overall score. These criteria were devised by an environmental expert specifically to tighten up this aspect of project selection within the SPD, basically bringing together the knowledge gained from experience of earlier programmes.

The second restrictive feature of implementation is that environmental monitoring (as with other aspects of physical monitoring) is still at an early stage of development. In Austria, existing environmental monitoring indicators are extremely general, but more exact and detailed monitoring is technically difficult and can be resource intensive; that is, the State government would not have the necessary time or resources to carry out more detailed monitoring checks, and firms would be equally unwilling. On the monitoring form, the question of assessing the environmental impacts of a project is also very general – more detailed questions were considered, but rejected on the basis of minimizing any additional administrative burden and complexity which might discourage companies from applying for support.

In France, the post-implementation assessment of impacts in the Aquitaine programme is considered to be straightforward for environmental projects, but only limited success has been achieved for projects in other areas; whereas in northern Sweden, implementation of the Ångermanlandskusten programme involves environmental monitoring of projects, but this originates from national legislative requirements, and in the Fyrstad programme, no specific monitoring arrangements have been made for environmental matters.

Lastly, in Finland, problems of monitoring include the difficulties of gathering data in the regions, particularly since each region may be covered by several offices. Nevertheless, the Environmental Forum in the City of Lahti Municipality has produced environmental indicators for monitoring over time. This project involves funding and collaboration between a number of public authorities, carrying elements of regional planning through to regional development strategy. Composed of around ninety participants and representing forty different organizations, the aim of the Forum is to adopt a common approach to the promotion of sustainable development in the Lahti area and to meet the challenge of Agenda 21. Its tasks include the development of local indicators for sustainable development to allow the evaluation of current conditions and the identification of specific objectives. The participants of the Forum will subsequently propose ways for different regional-policy actors to take these objectives into account.

Conclusions

In recent years, a range of initiatives has effectively raised environmental awareness internationally, and this has influenced perceptions and actions in both public and private-sector activities. The momentum initiated in the 1980s has been sustained through the 1990s with more sophisticated syntheses of environmental factors and public policy initiatives, and with the emergence and acknowledgement of ecological modernization, a direct and productive link has been identified between economic development and the environment. Correspondingly, within the institutions of the European Union, elements of a proactive approach towards environmental integration are apparent in the various environmental policy programmes launched by the European Commission and in progressive guidance for economic-development strategies.

Through the Structural Funds, the EU has been instrumental in promoting the widespread adoption of several principles within economic development including strategic planning, partnership, monitoring and evaluation, and environmental concern. Although clear progress has been made in environmental integration, in some cases there is still a tendency towards the rhetoric of sustainable development at the level of strategic objectives followed by the compartmentalization of environmental action within specific priorities or measures. This result is partly attributable to attitudinal factors, but also to institutional constraints and the traditionally weak position of environmental policy relative to economic, industrial or labour-market policies at EU and member-state levels.

Nevertheless, looking to the future, further positive developments can be expected to emerge. The outlook for the European Union envisaged in the Commission's Agenda 2000 document comprises both an expansion of the EU territory and a rationalization of elements of the Structural Funds, in particular reducing the number of objectives from seven to three and achieving greater financial efficiency. In parallel, a series of EU-generated evaluations is gathering information on scenarios in which the environmental impact of the Structural Funds can be identified, measured and directed to more positive ends. Against this background, opportunities can be identified to develop greater environmental integration within the post-1999 generation of EU regional development programmes.

First, during 1998 and 1999, many member-state authorities will be starting to plan the shape of new programmes. In northern European countries in particular, this programming is likely to become increasingly complex with multi-level planning, programme management and targeting of assistance, and national and regional-level authorities will have to consider the spatial interrelationships between the 'sub-regional elements' (urban, rural, industrial and fisheries) encompassed by Objective 2 as well as the localization of support within eligible areas. Throughout this preparation, strategic environmental direction and advice will be necessary from an early stage.

Second, transferable lessons could be identified from the experience of EU programmes. In terms of environmental progress, a number of programme partnerships are now developing new approaches to environmental integration, devising practical methodologies to assess the strategic-environmental elements of programmes, establishing indicators for environmental monitoring, and acknowledging the need for environmental categories within project-selection criteria. Other factors of interest relate to the successful use of special regional innovation strategies (RIS) and regional innovation and technology strategies (RITTS). Promoted by DG XIII and DG XVI of the European Commission, the RIS/RITTS initiatives recognize the difficulty that many regions have in promoting research and technological development and that the compressed programme planning process is driven primarily by regulatory and financial imperatives which afford insufficient time for full consideration of more challenging aspects of strategy development. There may be a useful parallel here with the issue of sustainable development, and programme managers may be able to benefit from exploiting similar dedicated 'regional environmental strategy' initiatives to enhance environmental integration within development strategies. Like RIS/ RITTS, this could involve the active exchange of experience and dissemination of best practice between regions utilizing forms of regional environmental-strategy support.

Lastly, greater emphasis and effort should be placed on environmental monitoring, identified earlier as one of the weaker aspects of environmental integration. Many aspects of the monitoring of physical outputs and impacts have yet to approach the efficiency and rigour of financial monitoring, mainly due to the methodological and institutional constraints of designing and applying suitable indicators and targets in difficult-to-measure areas of assistance. Nevertheless, the quality and detail of environmental monitoring can be expected to improve progressively within programmes, and as this occurs efforts should be made to expand the focus from environmental projects to encompass non-environmental projects. In addition to facilitating integration, such measures would draw together a much more comprehensive, insightful and influential data source on the positive and negative environmental impacts of economic development.

References

Bachtler, J. and Michie, R. (1994) 'Strengthening economic and social cohesion? The revision of the Structural Funds', *Regional Studies* 28(8).

CEC (1983) 'Third Environmental Action Programme 1982–1986', *Official Journal* C 46, 17 February.

CEC (1987) 'Fourth Environmental Action Programme 1987–1992', *Official Journal* C 328, 7 December.

CEC (1992a) 'Memorandum on the Current Environmental Dimension of the Revised Structural Fund Arrangements: A perspective with proposals for the future', DG XI, Brussels: Commission of the European Communities.

CEC (1992b) 'Court of Auditors Special Report No. 3/92 concerning the environment together with the Commission's replies', *Official Journal* C 245, 23 September.

CEC (1993a) 'Towards sustainability – A European Community programme of policy and action in relation to the environment and sustainable development', *Official Journal* C 138, 17 May 1993.

CEC (1993b) 'Growth, competitiveness and employment: The challenges and ways forward into the 21st century', White Paper, *Bulletin of the European Communities*, Supplement 6/93, Brussels.

CEC (1996a) *Progress Report from the Commission on the Implementation of the European Community Programme of Policy and Action in Relation to the Environment and Sustainable Development 'Towards Sustainability'*, COM (95) 624 final, Brussels: Commission of the European Communities.

CEC (1996b) *Draft Proposal for a Council Directive on the Assessment of the Effects of Certain Plans and Programmes of the Environment*, COM (96)511 final, Brussels: Commission of the European Communities.

Clement, K. (1994) 'Regional environmental policy: Dutch experiments with external integration', *European Environment* (4).

Clement, K. and Fitzgerald, R. (1997) '*Regional environmental integration: Changing perceptions and practice in Objective 2 programmes*', unpublished report, University of Strathclyde: European Policies Research Centre.

Gouldson, A. and Murphy, J. (1996) 'Ecological modernisation and the European Union', *Geoforum* 27(1).

Hildebrand, P. (1993) 'The European Community's environmental policy, 1957 to 1992', in D. Judge (ed.) *A Green Dimension for the European Community: Political Issues and Processes*, London: Frank Cass.

Johnson, S. and Corcelle, G. (1995) *The Environmental Policy of the European Communities*, second edition, The Hague: Kluwer Law International.

Larrue, C. (1994) 'Regional development and quality of the environment: The experience of OECD member countries' in B. Lindström, and A. Frovin (eds) *Regional Policies and the Environment*, Stockholm, Sweden: Nordrefo.

OECD (1990) *Experiences in the Implementation of Regional Development Policies in Areas Considered by Governments to Have Very Severe Problems*, Paris: Organisation for Economic Co-operation and Development.

Seamark, D. (1996) 'European funding and environmental appraisal', *Town & Country Planning* 65(12).

Therivel, R., Wilson, E., Thompson, S., Heaney, D. and Pritchard, D. (1992) *Strategic Environmental Assessment*, London: Earthscan.

Wilson, E. (1993) 'Strategic environmental assessment: Evaluating the impacts of European policies, plans and programmes, *European Environment* 3(2).

Woodford, J. (1991) *Conflict or Convergence? Environmental Priorities and the Structural Funds*, environmental policy discussion paper no. 1, University of Strathclyde: EPRC.

3 American perspectives on economic development and environmental management
Changing the federal–local balance

Martin Whitfield and Douglas Hart

Introduction

In America, economic development and environment management are not so much integrated with each other but are part of a continuously changing balance of forces which has been taking place for well over 100 years. An analogy which might be employed to illustrate this process is drawn from physics – Newton's Third Law of Motion – which states that, 'for every action there is an equal and opposite reaction'. Another way of looking at the relationship between the two is drawn from economics – taken together, environmental protection and economic development are commonly perceived to be a 'zero-sum game'. This concept means simply that more of one is purchased at the price of less of the other. According to this view it is possible to promote economic development, or to protect the environment, but not to do both simultaneously.

Extensive American experience of these two supposedly conflicting forces is instructive and demonstrates, among other things, that matters are far more complex than Newtonian Physics or the zero-sum-game analogy would suggest. Over the last century at least, in America one of the central issues has not been whether there will be economic development, or environmental protection, but where the balance will be struck between the two, by what means, and at whose expense. The economy and the environment operate more like a mathematical equation with each term affecting the other as part of a continuingly dynamic process.

The structure of government in the United States plays an important role in this dynamic. The US has a federal governmental system. At first glance, to those unfamiliar in any detail with federalism, it appears to be a simple hierarchy with national (federal) legislation made in Washington DC, and imposed on lower levels of government, rather in the way that unitary countries such as the United Kingdom operate with parliament passing laws which must be implemented by local authorities under the *ultra vires* principle. In fact the situation in America is not as simplistic as suggested, with independent local legislation being made by each of the fifty states and the local-authority jurisdictions within them, including tens of thousands of cities, counties and townships.

The role of the different levels of government, whether federal, state or local was defined by the Constitution of the United States and strengthened by the Tenth Amendment which declared that any powers not explicitly granted to the federal government were to be reserved for state or local authorities and ultimately for the people. Thus there is a reciprocal process: pressure emanating from the local level can (and constantly does) influence federal legislation and, as we have already indicated, at the same time federal legislation impacts on states and localities. As a consequence, environmental practice and patterns of economic development are extremely complex, diverse and dynamic in the US.

While it would be impossible in the space available in this chapter to chart this diversity in any detail, it is nevertheless possible to identify key themes which have emerged over time. This chapter presents an overview of the situation in the US by selecting and illustrating three of these themes: first, the changing balance over time between the supposedly unopposed economic exploitation of natural resources in Frontier America, with growing historical local and national environmental concern; second, and relatedly, the increasing role of the federal government in legislating mandatory local environmental-protection measures; and, finally, an assessment of the local impact of federal legislation on states and cities, with particular regard to the growing recognition of the costs of compliance with environmental-protection legislation not simply for business but increasingly for local communities as well.

Economic exploitation and the myth of the frontier

According to one widely-held view, America's exploitative frontier culture set the contemporary pattern for dynamic economic development in the US and elsewhere – and for environmental degradation. This view may be called 'The Myth of the American Frontier' and it suggests that economic exploitation occurred without any concern for environmental matters in Frontier America. It is a myth because it is not so much completely untrue as a partial and a simplified exaggeration about what was actually taking place in the US even at the end of the nineteenth and the beginning of the twentieth century. As we will demonstrate later in this section of the chapter, concern with environmental issues, even during this period, was more in evidence than the largely economically oriented myth of the frontier would suggest.

The myth runs like this: frontier culture encouraged reckless resource exploitation, often without regard to either conservation or environmental impacts (Garvey 1972). The apparent limitless natural bounty of the North American continent engendered heedless, wasteful habits as the early frontiersmen and their latter-day successors appropriated land, forests and water resources. Crucial to this belief was the view that there would always be more of these natural resources further west as the frontier advanced and that any negative impact that occurred to the environment could be safely ignored. Local material prosperity was satisfied by relentless resource plundering of natural assets by the early hunters and farmers, and subsequently industrial entrepreneurs, as the process of taming over time moved from initial exploitation to ultimate degradation of whole forests, lakes and rivers.

Eventually, this pattern of expansion was also carried over into the fields of energy – particularly coal and petroleum development. The energy entrepreneurs who succeeded the early farmers and hunters operated on a much larger scale than their predecessors and were more concerned with short-term profits than long-term environmental impacts. Following the examples of the forests, the Appalachian coal fields were rapidly degraded. Next came the development of oil and gas in the Mid-West, and ultimately the Far-West, with the early drilling fields being so intensely farmed that one oil rig literally overlapped another.

According to the frontier myth, with nature apparently having such inexhaustible bounty on an entire continent, there seemed little reason to regulate its exploitation, indeed during much of this period the role of the federal government was far more that of a facilitator of economic expansion, not as an environmental guardian. The economic doctrine of *laissez-faire* with the federal government – and abated by state and local government – adopting a 'hands-off' approach received much favour in America during the late nineteenth and early twentieth century.

In the US, in an early but massive example of privatization, natural resources were passed from public to private domain at an unprecedented scale, initially through land, followed by valuable minerals and finally into the air and water. This occurred to the extent that on land whole forests disappeared, and waterways were polluted – even the huge inland seas – the Great Lakes – began to die as a result of being used as vast sinks to absorb wastes created in pursuit of private profit. Paraphrasing the economist John Kenneth Galbraith, the situation was one of private greed and environmental squalor.

The myth suggested that in the United States there has been a long history of unopposed environmental degradation, much of it due to the economic development that pushed back both the economic and libertarian frontiers and promoted individual materialism. Whether because of the resources economic development extracts; the processes it involves; or the products it manufactures, distributes, consumes and ultimately discards, economic activity has been a major contributor to environmental degradation (Welford and Gouldson 1993).

All of this is true, but even at the time that this degradation was taking place, matters were more complex than the frontier myth seemed to suggest. Even during the height of *laissez-faire* economic development at the end of the nineteenth century, individuals and local groups in America expressed concern about the impact of such developments on the environment. This realization occurred particularly during the last decade of the 1800s. For example, in his 1890 report the director of the US Census Board expressed the simple but important view that 'the frontier had now disappeared'.

There was a growing recognition of the finite limits even to the vast American continent. Many in government accepted this view and its implications and confirmed a long-term concern which some had been expressing for some time. The worry might be summarized as, 'if we continue to use our finite resources for economic development without heed to the future, then we will inevitably exhaust them – and impoverish ourselves and our children in the process'. Far-sighted people could see that the official acknowledgement of the end of the frontier also meant the end of the frontier myth. Clearly conservation of existing resources was required to safeguard the future.

The development of preservation and conservation measures and the advent of active environmental protection

One year after the statement that the frontier had vanished, the first significant piece of federal environmental legislation was passed: The General Revision Act (1891), which contained among its regulations permission for the President to set aside 50 million acres of timberland as national forest. By 1907 the national forests had increased to 150 million acres which were administered by the newly formed United States Forest Service. There were 75 million acres of federal land which was reported to contain valuable minerals and proposals were established for water-resources management schemes. This period was described by a presidential aide as being 'the first conservation movement' and this early activity also saw the establishment at both the state and national level of the Sierra Club in California, and the National Audubon Society in Washington, DC. The pressure for change stemmed initially from those who were concerned that the economic development of their forefathers through many decades of frantic exploitation was likely to result not only in despoiling the environment but in the exhaustion of resources upon which they were now reliant. Thus the early conservation movement was not simply concerned with conserving nature but with the preservation of economic self-interest.

The first major wave of concern about the long-term consequences of unfettered local economic development subsided during the two decades 1910–30 – largely due to the First World War and its aftermath (O'Riordan 1971). After World War I, America faced environmental disasters such as flooding and the Dust Bowl in the Mid-West – which were at least partly man-made – and which resulted in a second wave of environmental commitment in the 1930s, particularly after 1933. It was in this year the Franklin D. Roosevelt administration, under the 'New Deal', emphasized the alleviation of resource concerns such as soil conservation.

Again these efforts were side-tracked by the Second World War, but during the 30-year period between the end of World War II and the beginning of the 1970s, there developed a set of local concerns which broadened environmental awareness beyond those addressed during the earlier conservation and preservation era. These concerns emphasized the link between people and their natural surroundings and thus tied environmental concerns not simply with economic self-interest but with the quality of everyday life. Gradually concern shifted from natural-resource conservation to active environmental protection.

Occasionally, one important work can radically change perceptions and awareness. The early 1960s saw what is generally considered by environmental historians as the birth of modern mass environmental activism in the US, with the publication of Rachel Carson's book *Silent Spring* in 1962 (Adler 1995; Sale 1993; Worster 1993). Carson was a biologist by training and demonstrated how the considerable post-World War II increase in the application of pesticides posed a significant threat to the global ecosystem. Carson named the economic activities she felt were responsible and in return the pesticide and petrochemical industries attacked her book.

Silent Spring introduced newer, and wider concerns, such as anxieties over chemical pesticides including DDT, and radioactive poisons with very long half-lives such as strontium-90, released by nuclear testing, that demanded a much broader approach than existing conservation measures *per se* gave. These were environmental concerns which tended to: (a) have delayed, complex and difficult-to-detect effects; (b) have consequences for human health and well-being as well as the natural environment; and (c) were more complex in origin, often stemming from new technologies (Mitchell 1989). Combining these new concerns with the revival of concerns over the long-term supply of natural resources, which appeared again in the late 1960s brought about the view that they should be seen as threats to the quality of life (Hays 1987).

Although older waterborne diseases were now largely controlled, there was rapid accumulation of newer chemical pollutants in the water system, such as synthetic organic compounds, and heavy metals from industry were discovered in the nation's drinking water. Toxic-waste disposal also became a pervasive concern due to the potential to seep into the water supply. Societal concerns were now firmly focused on local economic activity, and in particular to the means by which industry, on occasion, was degrading the environment and threatening human health (Ashford 1977; Berman 1978).

As a consequence, the 1960s saw a massive rise in the number of federal legislative bills being passed, with over twenty major pieces of legislation including the Wetlands Act (1961), the Federal Water Pollution Control Act (1961), the Solid Waste Disposal Act (1965) and the Clean Air Act (1967). Many were directed at, or contained elements aimed at, controlling the types of local economic development that created the most severe environmental degradation.

But environmental concerns were not limited to federal legislation. Continuing local-level concerns gave rise to the first Earth Day in 1970 which demonstrated a growing awareness of the global nature of environmentalism. This event was promoted as the beginning of contemporary mass environmentalism, with between 10 and 20 million people across America participating in rallies and other activities over 2,000 campuses and local communities, along with 10,000 high schools. Pressure for environmental change had reached new heights. For the decade that followed this first Earth Day pressure to change individual, corporate and governmental behaviour towards the environment produced some success. Unfortunately these successes were not without their failures.

One of the most important failures was the lack of any meaningful penalties for firms which failed to comply with the federal legislation. Although more than half of the 439 federal pre-emption statutes passed by Congress in the 200-year history of the United States were enacted in the last two decades (Pompili 1995), those introduced throughout the 1950s and 1960s were seen to have very little in the way of non-compliance penalties and were thus relegated to state or local agencies to implement with only minimal federal oversight (Davies and Davies 1975). Indeed, as we will demonstrate later in the next section of the chapter, the cost of compliance, and the burden on whom the cost fell, was to become one of the main issues for environmentalism in the late 1980s and 1990s.

Returning to our historical narrative, during the 1970s, as local pressure for change expanded, concerns about environmental quality were expressed not only to local companies and local government but pressure was also extended upwards through state authorities and ultimately to the federal government. During the 1970s, the federal government of the United States played an increasingly significant role in improving the quality of the US environment and this improvement has continued – with some significant opposition, and some notable exceptions – over the last two decades.

On the first day of the 1970s, for example, the incoming President, Richard Nixon, introduced the National Environmental Protection Act (NEPA) which mandated every federal agency to protect the environment. Further legislation followed, with two landmark bills: the Clean Air Act (1970) and the Water Pollution Control Act (1972). When Richard Nixon signed the National Environmental Protection Act, he stated, 'The 1970s must absolutely be the years when America pays its debt to the past by reclaiming the purity of its air, its water and our living environment. It is literally now or never.' In the early 1970s legislation was used by federal government to impose more control over state and local governments, with additional environmental controls increasing drastically (Pompili 1995).

The legislation introduced was crafted by politicians and environmentalists working together in order to reduce environmental degradation and restore air, water and soil quality. Largely as a result of this partnership over 10 million acres of land has been added to the federal wilderness programme, environmental impact assessment is now a prerequisite for all major developments, lakes once described as dead show signs of life and although not pure, the air and water quality which US citizens now experience is certainly less toxic today than would have been if no such legislation existed – and in some cases pollution has actually been reduced and environmental degradation reversed.

In assessing the impact of this body of legislation, in 1989 the President's Council on Environmental Quality stated:

> Perhaps the greatest progress has been made in controlling air and water pollution where concentrations of many pollutants are showing measurable decline. Emissions of total suspended particulate, sulphur dioxide, nitrogen oxides, volatile organic compounds, carbon monoxide, and lead from various sources have been reduced in the past decade as a result of pollution controls.

Table 1 National standards for the six primary air pollutants

Pollutant	Emissions* 1970	Emissions* 1990	Percentage change
Particulate matter	18.5	7.4	− 60.0
Sulphur oxide	28.3	20.7	− 26.9
Nitrogen oxides	18.5	18.8	+ 1.6
Volatile organic compounds	25.0	16.9	− 32.4
Carbon monoxides	101.4	62.1	− 38.8
Lead	203.8	5.0	− 97.5

Source: *Statistical Abstract of the United States* 1993.

* In millions of metric tons, except lead in thousands of metric tons.

Concentrations of suspended solids, oxygen-demanding wastes, and phosphorous are declining in many waterways. There has been a marked reduction in environmental levels of DDT and other persistent organochlorine pesticides; polychlorinated biphenyls (PCBs); vinyl chloride; benzene; asbestos; and mercury, lead and other heavy metals. Concentrations of these and other chemicals in human and wildlife tissues have also declined.

In quantitative terms, the effectiveness of the Clear Air Act can be demonstrated over a 20-year period. The impact of the act is illustrated in Table 1. It should also be noted that, as part of the century-old dynamic tension between environmental protection and economic development, at the same time that new legislation protecting the environment was being implemented, there was a growing counter-offensive from corporations against the cost of the environmental successes to the private sector, and this offensive was increasingly supported by the White House during the late 1970s and 1980s. Pressure for governmental change towards more environmentally sensitive policies and practices were consequently reduced.

But the century-old persistent tension between the local and the federal level continued unabated. And even as federal support for environmental change waned during the second Reagan administration and the Bush presidency, so grassroots environmentalism found new life and many local environmental pressure groups found their memberships expanding rapidly during the 1980s and early 1990s. To take just one example, the influential Wilderness Society had 68,000 members in 1981, this had grown to 350,000 by 1992 (Dowie 1992). Thus pressures on local government were not only being felt as a result of mandatory federal legislation, but also because of local discretionary influences, such as community-based and even national environmental pressure groups.

As we have already indicated in our brief historical review, this dynamism and diversity with regard to the economic-environmental balance are not new. What is new, is the pace, scale and scope of the changes – and their cost. During the

1970s the total number of state and federal toxic-management mandates was eleven. Between 1980 and 1985 a further nine more mandates were imposed on local government and in the 4 years up to 1992 a further seventy-five toxic-management mandates have been imposed upon local government. Some of these seventy-five have implementation dates up to the year 2015 (Pompili 1995). The rate at which legislative controls have been introduced in the 1990s has not slowed. The Semi-Annual Regulatory Agenda, which is published in the Federal Register, showed that between October 1990 and April 1994, 2,720 further actions were introduced.

Pressure is thus being exerted by grassroots environmentalists on all levels of government. In response to this pressure and global concerns federal government has increased, and will continue to introduce, environmental mandates which necessitate local government implementation. As one observer noted:

> To many public officials, and to many observers, the reign of what had been termed co-operative federalism seemed to be at an end. Other 'c' words – like compulsory, coercive, and conflictual – were suggested to describe patterns of federal-state-local relationships that centered not on well-established financial grants-in-aid programs but on disputes over new kinds of federal rules and regulations.
>
> (Beam 1990: 23)

As a consequence, local government is caught in the middle – it is under significant and growing pressure both from the people it represents and the nation's government. The overall effect of such pressure was to seriously affect the ability of such local governments to remain in complete compliance with federal legislation while at the same time providing essential local services within their existing budgets.

The impacts on local government and local communities: The cost of compliance

Throughout the 1970s and early 1980s the impact of these new mandates experienced locally by both the authority and the taxpayers it represented was minimal. A significant proportion of the extra costs which implementing mandates created was reimbursed from federal funds, although no reimbursement was forthcoming for the extra costs borne by the private sector. This reimbursement took the form of financial incentives or grants, such as payments made for sewer and water projects. The mid-1980s, however, saw these grants reduced, eliminated or converted into loans due to pressure on the federal government to achieve a balanced budget. Many of these loans were offered to local government not by federal but state authorities. In 1990 the US Environmental Protection Agency (EPA) highlighted the growing problem which faced local governments. In 1981 it was seen that $35 billion was needed if the 1987 environmental quality standards were to be achieved. Of this, local authorities were expect to fund 76 per cent, with the state providing 6 per cent and the EPA 18 per cent. By 1987 the public expenditure figure had risen to $40 billion in order to maintain standards, of which 82 per cent was provided locally, 5 per cent from state funds and 13 per cent from the EPA. It is predicted that by the year 2000, $55 billion will need to be expended if environmental quality is to be maintained at 1987 standards. Of this, the EPA is predicted to provide just 8 per cent, with the state providing 5 per cent and 87 per cent is to be sourced from the local community. The financial relationship between, and among, the various levels of government in America has changed considerably over the past decade with the incidence of expenditure continuingly being pushed down to the local government level.

As this funding shift was occurring, so another intergovernmental relationship was being restructured. The resources that federal government provided had additional conditions attached to their allocation, not only in the environmental arena, but in the majority of federal allocations made to state as well as local authorities. The overall effect was to decrease the level of funding for environmental causes, reallocating these resources towards other governmental concerns and attaching additional controls and costs.

The combined impact of increasing the number of mandates imposed on local government which aimed to integrate economic development with environmental management, along with the decrease in financial support provided by the federal government for their successful implementation, has led to pressure being placed upon local communities, pressure which could have devastating effects on many smaller communities.

Even though the relationship between the federal and local government has changed considerably since the mid-1980s, it is the latter which remains the central focus for the provision of services which are not only timely and cost-effective but remain in compliance with established and newly passed environmental legislation. In order to achieve this it is essential that the financial cost of compliance is established. Many communities across the US are undertaking such studies, including the city of Columbus, Michigan which was the first to produce a 'Cost of Compliance' report. The report, entitled 'Environmental Legislation: The Increasing Costs of Regulatory Compliance to the City of Columbus' was published in May 1991. The findings showed that an additional $1 billion would be required if compliance was to be achieved with all environmental mandates enacted as of January 1991. A decade later, the local authority predicts an increase in taxation for each individual household to the extent of $856 per annum. Associated with this was the fact that overall financial resources would be limited, less would be available for other community services and authorities would have less freedom with respect to the allocation of resources. Essentially local government has experienced a shortage in the level of funding it receives to integrate economic development with environmental management. This has necessitated many authorities to turn to their own taxpayers for the additional funding, in many cases it is the very same taxpayers who are demanding environmental improvements. This has been the case across the US. In

Hastings, Nebraska (population 22,867) the total cost of compliance is predicted at $74.6 million, at a 'per household level' this is reflected as an additional $1,865 per annum for each member of the individual household. In Anchorage, Alaska the $429,936 million total cost of compliance can be expressed as an additional 10-year cumulative cost per household of $4,659 (Pompili 1995).

City-cost studies are not the only reports which show seriousness of the problem for local governments and their communities. The Association of Metropolitan Sewers Agencies (AMSA) issued its report 'The Cost of Clean' in June 1992. Surveying 108 metropolitan areas of the US it determined that municipalities would be required to raise $22.6 billion in the first half of the 1990s simply to cover mandated capital improvements. Only $1.8 billion was to be provided from federal funds, which is less than 8 per cent of the total cost. The burden upon the local authorities is thus immense. Establishing a cost-per-household over 20 years the report stated the average annual household-user fee will double every 6 years to the cost of $1,695 per annum/per household by the year 2010. The financial implications of integrating economic and environmental concerns for both the local governments and the local communities are obvious, but there is a second impact that this integration has created, that of social impact.

It is now apparent that the cost of compliance has increasingly had a physical as well as a financial cost. By the early 1980s the anti-toxic movement and environmental-justice movement became known as the NIMBY movement (Not In My Back Yard). The NIMBY movement protested against the locating of hazardous-waste processing and dumping facilities within communities. Many of these protests were opposed to the locating of such facilities in white middle-class areas. The combination of this local opposition with the increasing number of toxic-management mandates being passed down from federal to local government resulted in many public officials and private industries responding with the PIBBY principle (Place In Blacks' Back Yards). In the early 1980s anti-toxic organizers in black and other racial-minority communities had initiated protests against the disproportionate number of toxic-waste sitings within their communities. In 1982, for example, anger and frustration within the African–American community of Warren County, NC over the siting of a PCB landfill site in their community resulted in the largest civil-rights protest since the 1960s (Alston 1990). Other protests followed; for example, the predominantly Mexican–American Mothers of East Los Angeles successfully blocked the establishment of a' toxic-waste incinerator, and the Western Shoshone Nation protested against the proposed nuclear-waste repository at Yucca Mountain, NV (Darnovsky 1992). Race, even more so than class, was the determining factor in the location of hazardous-waste facilities (Chavis and Lee 1987; Bullard 1990) and this practice was termed 'environmental racism' by the Commission for Racial Justice.

The changing balance between economic development and environmental management has brought with it the benefits of at least a partial reduction in the level of environmental degradation. However, in many instances these environmental benefits have been associated with negative economic impacts.

Financial resources are being squeezed which has affected the provision of community-based services and it has been seen that the pressure placed on local-government officials to satisfy both federal and local demands has resulted, in some cases, in certain elements of the local community experiencing what might be called environmental prejudices. What recommendation can be made for the future role of local government in seeking to redress the balance between economic development and environmental protection?

Recommendations

Local governments should be aware that the changing relationship between them, the state and federal government will have a serious effect on their ability to provide the wide range of community services being demanded. Whilst at the same time the achievement of the economic and environmental demands that are being expressed both locally and nationally will become increasingly difficult. Those individuals who take the decisions within local government should educate themselves as to the dynamics of this changing relationship so that corrective action can be implemented. It is never too soon to introduce such policies.

Local governments should accept that in the US the taxpayer of the 1990s may hold different values than those of the 1980s. They demand their local environment and community is not polluted; furthermore, they demand that the dollars collected via taxation be used efficiently and effectively. Transparency within local government should be paramount, taxpayers demand the ability to see where their money is going, in essence taxpayers demand accountability. They are also realizing the link between undertaking unfunded federal mandates and the level of community-service provision. The local citizenry are actively taking part in the local priority-setting practice, weighing the benefits of undertaking unfunded federal mandates with that of reducing services elsewhere. The simultaneous expression of environmental and economic concerns has resulted in an increase in local-community involvement within local-government decision-making processes. This involvement should be actively encouraged as it provides the opportunity for local citizenry to establish the priorities they deem necessary for their own well-being. It also provides the opportunity to local government to explain why a rise in tax rates is required, and the expected benefits such an increase in financial resources should bring. Without this full co-operation, the local government is unlikely to maintain continued support which in turn may breed resentment among its citizenry and further opposition to other local initiatives whether environmentally related or not.

The process which permits an increasing and cumulative mandate burden to be placed on local government whilst also restricting the financial resources provided from federal reserves should be addressed. One solution has been suggested by Pompili (1995) whereby the role of the federal government should include, but should not be limited to: funding for environmental research; the establishment of common environmental indicators; appropriate tools to measure success and/ or failure (not mandating specific compliance levels for all communities);

matching funds for initiatives where significant funding is required; training, educational and technical resources to other government entities; and inter-national leadership on world-wide environmental concerns. If federal government devoted their efforts to accomplishing these roles, this would permit state and local governments to perform the role they are best suited to, that of meeting the demands of the local communities through the re-direction of resources. This is a move away from the 'one size fits all' approach that is commonplace in the US today.

Two recommendations have been made. One affects the relationship between the local government and its citizenry by actively involving them in the prioritizing of local issues. The other affects the relationship between local and the federal governments by clearly defining the role each should undertake, but both aim to reconcile the conflict between the desire to promote both economic development and environmental management and the need to provide adequate, timely and efficient local-community services.

Conclusion

The United States has a long history of seeking to balance conflicting demands on resources. Unprecedented economic growth grew hand in hand with local environmental degradation. The vast majority of this degradation was the consequence of the economic activity that created today's nation in which, as President Calvin Coolidge stated, 'The business of America is business'. Economic activity is the central tenet of American society, but since the turn of the century the US people have become increasingly concerned with the costs, as well as the benefits, of this activity. Initially established as a conservation movement today's environmentalism grew from the influential work of authors and scientists such as Rachel Carson. Her book *Silent Spring* broadened the environmental debate to include quality-of-life factors, which through the decades that followed increasingly focused on the negative aspects of economic activity. Over the last 30 years the US government has made a concerted effort via legislation to balance economic development with environmental management. Prior to the 1970s most federal legislation introduced lacked any real means of punishing non-compliance.

At the same time as this increase in mandatory federal legislation was occurring, there was a decreasing allocation of federal financial resources. Pressure on local government to improve the quality of the local environment was also being exerted via grassroots environmentalism including the NIMBY movement. Local government found itself being called both to 'Think Globally' and 'Act Locally', experiencing top-down pressure for change from the federal level above; at the same time it was experiencing bottom-up pressure from its citizens below. It also found itself with minimal resources available to achieve its objectives.

The result has been that local authorities must look towards the very people that are demanding environmental improvements, the taxpayers of the local community, both as a source of ligitimacy and as a source of funds. Unfortunately before an increase in financial resources can be achieved from this source, the taxpayers must be convinced that the existing level of financial resources raised via taxation is being used appropriately and efficiently, only then will they permit more of their dollars to be removed. In order to satisfy taxpayers' demands for accountability local authorities are increasingly required to improve their transparency. They are increasingly required to open up the decision-making process to public scrutiny.

This is one of the recommendations proposed earlier: in order to accelerate the integration of economic and environmental concerns the very public that such activity affects must be able to witness and agree the resultant effect upon the local community. The second recommendation is the restructuring of the roles played by each level of government. The result of this restructuring should be an increase in the flexibility within local government which will enable maximum environmental benefits from the increasingly limited financial resources available. The increase in federally imposed mandates while at the same time decreasing resources available for their implementation cannot continue unchecked.

The United States has made significant steps towards balancing the forces of economic development with environmental management, particularly within local government, and in some instances this has resulted in substantial improvements in environmental quality. However, it is becoming increasingly apparent to local authorities, as it has been for sometime to local businesses, that the environment itself is not a free good.

References

Adler, J. (1995) *Environmentalism at a Crossroads*, Washington, DC: Capital Research Center.

Alston, D. (ed.) (1990) *We Speak For Ourselves: Social Justice, Race and Environment*, Panos Institute: Washington, DC., pp. 7–8.

Ashford, N. A. (1977) *Crisis in the Workplace*, Cambridge, Mass.: MIT Press.

Beam, D. (1990) 'On the origins of the mandate issue', in M. Fix and D. Kenyon (eds) *Coping with Mandates*, Washington, DC: The Urban Institute Press, pp. 23–31.

Berman, D. M. (1978) *Death on the Job*, New York: Monthly Review Press.

Bullard, R. D. (1990) *Dumping in Dixie: Race, Class and Environmental Quality*, San Francisco: Westview Press.

Carson, R. (1962) *Silent Spring*, New York: Fawcett Crest.

Chavis, B. F. and Lee, C. (1987) *Toxic Wastes and Race in the United States*, New York: United Church of Christ Commission for Racial Justice.

Davies, C. J. and Davies, B. (1975) *The Politics of Population*, Indianapolis, IN: Bobbs-Merrill.

Darnovsky, M. (1992) 'Stories less told; histories of US environmentalism, *Socialist Review* 22(2), pp. 11–54.

Dowie, M. (1992) 'American environmentalism: A movement courting irrelevance', *World Policy Journal* 9(1), pp. 67–92.

Garvey, G. (1972) *Energy, Ecology, Economy; A Framework for Environmental Policy*, London: The Macmillan Press.

Hays, S. P. (1987) *Health, Beauty and Permanence: Environmental Politics in the United States 1955–1985*, New York: Cambridge University Press.

Mitchell, R. C. (1989) 'From conservation to environmental movement; The development of the modern environmental lobbies', in J. M. Lacy (ed.) *Government and Environmental Policies*, Washington, DC: Wilson Center Press, pp. 81–113.

O'Riordan, T. (1971) 'The Third American Conservation Movement; New implication for public policy', *Journal of American Studies*, 5, 155–71.

Pompili, M. J. (1995) 'Environmental mandates: The impact on local government', *Journal of Environmental Health*, 57(6), 6–12.

Sale, K. (1993) *The Green Revolution*, New York: Hill and Wang.

Welford, R. and Gouldson, A. (1993) *Environmental Management & Business Strategy*, London: Longman Group.

Worster, D. (1993) *The Wealth of Nature: Environmental History and the Ecological Imagination*, New York: Oxford University.

4 Approaches to the integration of environmental protection and economic development in local government in New Zealand

Jennifer Dixon and Neil Ericksen

Introduction

For over a decade, the New Zealand government has been pursuing a programme of reform and restructuring. The rhetoric of the reform process has been to get 'government out of business and business into government'. However, the reform process has also been about reducing central-government influence in local-government affairs, thereby making councils at the regional, city and district levels more directly responsible for the problems experienced at the local level. For councils, these changes were enacted in amendments to the Local Government Act in 1989 and through the adoption of the Resource Management Act in 1991. These acts provide the current statutory basis for councils to deal with issues on environment and development (May *et al.* 1996).

This chapter begins by summarizing the salient features of the recent reforms and by introducing the relevant processes in local government that relate to economic development and environmental management. The chapter then explores the extent to which an integration of economic and environmental policies and objectives is occurring under the new institutional arrangements. Various planning instruments which have been created as a consequence of the reforms are considered and some of the current issues related to implementation are discussed. Examples of tensions between environment and development occurring at local and national levels are assessed. The chapter concludes with some observations about the dissonance between the intentions of policies designed to promote an integration of environmental and economic objectives and the realities of policy implementation in the devolved and co-operative system that is now in place.

Nature of recent reforms

For 50 years, New Zealand developed under a mixed economy that was designed to promote not only economic development, but also social welfare. Largely as a consequence of state-sponsored public-works programmes, national debates over economic development and environmental protection escalated in the 1960s and intensified until more lasting solutions were provided in the

1980s. The mid-1980s saw the beginning of a programme of wide-ranging reforms to central and local (regional, county and municipal) government driven by new-right ideology that sought to promote market-led, user-pays strategies and the devolution of decision-making to local levels. The latter aimed to make local councils responsible for the identification and application of policy responses to local problems.

In searching for strategies that gave due weight to economic development and environmental protection at the local level, central government emphasized 'sustainable management' over 'sustainable development'. The result of this has been the separation of the environmental management and economic development functions of regional and local (i.e. district and city) councils. Consequently, regional councils are now responsible for the integrated management of natural and physical resources while local councils are responsible for land-use planning, subdivision, service delivery, etc. While regional councils now manage natural resources such as water and soil, local councils carry out multiple functions which enable them, amongst other things, to promote development projects with the private sector. Local councils are not, however, involved in educational, health and welfare matters which remain in the domain of central government (Howell *et al.* 1996).

The restructuring of local government

The programme of central-government reform was driven by the principles of transparency and accountability and the provision of 'client-driven' services. These principles were also applied to the programme of local-government reform. Some 234 local authorities were reduced to seventy-four local councils including fifty-four district, fifteen city, four unitary and one county council. In addition, forty-two united/regional councils and catchment/water boards and a host of *ad hoc* authorities (like harbour boards, pest-destruction boards, noxious-plant authorities, and river and drainage boards) were reconstituted as twelve regional councils. With appropriate citizen support, councils can argue for unitary authority status which provides the dual functions of regional and local councils. There are currently four unitary authorities. While local councils are situated within the spatial boundaries of regional councils, local councils are not subordinate to regional councils. Rather, the regional council is supposed to be in complementary partnership with its local councils. Thus, the regional council and local councils share in the governing of their areas.

All councils have corporate status and can exercise a degree of autonomy within the areas for which they have been granted competence. However, their activities are also influenced by other key elements of local-government reform, relating for example to the separation of political and managerial functions from service delivery and regulatory functions. Councils are also obliged to prepare annual plans, to consult with the public on these plans, to establish measures of performance based on required outputs and to publish annual reports on their activities. The reform process, which is on-going, has also introduced an obligation to consider corporatizing the trading activities of regional and local councils, for example through LATEs (Local Authority Trading Enterprises), business units and mixed private sector and in-house arrangements, if not through complete privatization.

Resource-management-law reform

The amalgamation of the body of resource-management law, which constituted some fifty-nine statutes and nineteen regulations and orders, resulted in the introduction of the Resource Management Act of 1991. The single purpose of this act is the sustainable management of natural and physical resources. It is distinguished from its predecessor, the 1977 Town and Country Planning Act, in two main ways. First, the new act emphasizes effects-based planning, whereas the old act emphasized activities-based planning. Second, the new act gives much greater precedence to the environment than the old act that functionally embraced development, economy and the environment.

In promoting the Resource Management Act's single purpose of sustainable management of natural and physical resources, regional and local councils must manage the use, development and protection of these resources in ways that enable people and communities to provide for their social, economic and cultural well-being and their health and safety while at the same time achieving three environmental objectives. These objectives seek to ensure that the life-supporting capacity of air, water, soil and ecosystems is safe-guarded, that the adverse environmental effects of activities on the environment are avoided, remedied or mitigated and that resources are conserved to meet the needs of future generations.

Purposes and functions of local government

The purposes and functions of local government are therefore embedded in two main statutes. The Local Government Act lists nine purposes which assert that 'New Zealand is made up of differing communities with differing needs and that local government provides a means whereby those affected can actively determine the nature and meeting of those needs' (Bush 1995: 125–6). To these purposes are added the need for efficiency, effectiveness and competitive neutrality. On the other hand, the Resource Management Act requires councils to operate in ways that meet its single purpose, namely the sustainable management of natural and physical resources in their areas.

Following restructuring the new regional councils absorbed most of the functions of their predecessor agencies. Functionally, the new regional councils are confined to regulating and managing the natural environment in an integrated manner that takes account of the quality of natural resources. They also manage the regional aspects of civil defence and transport planning. To achieve these functions, they must produce publicly approved regional-policy statements, transport strategies and coastal plans and they may prepare other regional plans.

Complementing the activities of regional councils, local councils have a wide range of functions. Some are mandatory, like environmental management, local aspects of civil defence and building and public health inspection, but most are discretionary. As Bush (1995: 127) notes:

> at least three *intra vires* powers confer on TLS [territorial local authorities; i.e. district and city councils] potentially enormous latitude. They can: (1) carry out works for the benefit of the district; (2) widely involve themselves in community development and welfare; and (3) impose bylaws for the good rule and government of the district.

Unlike regional councils, local councils can engage in the promotion of economic development, although prevailing new-right attitudes suggest that this should be the function of the private sector rather than of local councils (New Zealand Business Roundtable 1995). Local councils may, however, still foster community self-determination, for example, through the provision of seed capital, subsidized salaries and the provision of facilities. They can also stimulate economic activities such as land and building-development programmes aimed at industrial or commercial use and encourage the improved economic and employment performance of private enterprise through such means as establishing resource centres, developing on-line databases, leasing buildings to small businesses, co-ordinating business and industrial development and reducing rates for manufacturers and industrialists (Bush 1995: 146–7).

Local councils are, however, required by the Resource Management Act to engage in environmental management, including land-use management, subdivision consents, noise control and natural and technological hazards. To achieve the purpose of the act, local councils are required to prepare district plans which deal primarily with these matters.

Processes and procedures for local government

There are several mandatory and non-mandatory processes which local government can enact to promote economic development and environmental management, some of which may be construed as enabling their integration.

Regulatory instruments

Under the Resource Management Act, a hierarchy of policies and plans is required which must not be inconsistent with one another. These instruments include: a national coastal policy statement to be prepared by the Department of Conservation for its Minister; regional policy statements and regional coastal plans to be prepared by regional councils; and district plans to be prepared by district and city councils. The act enables, but does not require, other national policies and regulations and regional plans to be prepared. Regional policy statements are based on publicly identified resource-management issues and contain a cascade of interrelated objectives and policies for achieving integrated management of natural and physical resources in the region. Regional and district plans do likewise, but go further in that they include rules or methods for achieving specified objectives.

Once councils have established policies and rules for sustainably managing their environments, they must deal systematically with resource developers who by law have to apply to the councils for resource consents or permits for development approvals. Developers, big and small, public and private, must show that their proposed activities will not have adverse environmental effects. In approving developments that are compatible with environmental objectives, councils thereby facilitate economic and social development in their areas.

Incremental, micro-level land-use changes that are environmentally compatible ought to result in good environmental outcomes. This, however, cannot be taken for granted. Accordingly, the Resource Management Act requires councils to monitor resource-consent compliance, the effectiveness of policies and plans and the state of the environment. There are also provisions for ensuring compliance, such as enforcement orders and abatement notices.

Amendments to the Local Government Act established an obligation for all councils to embark on a publicly notified annual planning and reporting cycle. A central part of this process is the preparation of an annual plan and annual report on performance. The annual plan should identify the significant objectives, policies and activities of council. It should also forecast annual expenditure on the proposed activities and indicate funding sources, including the policy on local rates and taxes. Further amendments to the Local Government Act require each council to produce a Strategic Financial Plan that looks 10 years ahead. These plans should be prepared every 3 years to show how funding in relation to activities may be achieved in the long term. Obviously, activities conducted under the Resource Management Act, such as monitoring resource consents, plans and environmental outcomes and undertaking implementation strategies are included as items that require funding in these annual and financial plans.

Non-mandatory initiatives

The resource-management policies and plans described above are considered to be only one of the mandated means for achieving environmental outcomes and councils are encouraged to use other measures where appropriate and desirable. These include the provision of public information and education programmes, the application of economic instruments such as transferable water permits and transferable development rights, measures to promote the development and diffusion of new technologies and techniques and the introduction of environmental standards that could be achieved by other non-regulatory means (May *et al.* 1996). In addition, a mix of non-regulatory instruments may be used by local government to encourage and plan for public infrastructure to support economic and social development. Related measures include the preparation of strategic plans in non-mandatory areas, the formulation of structure plans, joint venture arrangements with the private sector and rates rebates.

The extent to which these non-mandatory initiatives are used to promote economic development or environmental management is determined by the councils themselves. Councils are involved extensively in a range of social, cultural and economic activities. These relate, for example, to the provision of recreation and sports facilities, parks, housing, libraries, museums and environmental health (Howell *et al.* 1996). A few councils have become involved in major initiatives with the private sector. Several councils have also engaged in the Agenda 21 Model Communities Programme (Ministry for the Environment 1994). Thus, there is considerable variation in the contribution of councils throughout New Zealand to social infrastructure and community development.

An integrated system?

The purposes, functions and methods given to local government under the Local Government Act and the Resource Management Act may be seen as elements of an integrated system. The main components of which are (1) resources; (2) policies and plans and (3) sustainable development (Figure 1). In practice, a closer look reveals that a number of factors impede efforts to integrate environmental management and economic development. These are: the integration of planning instruments; implementation issues; local tensions; and concerns of national corporates. Each of these factors will now be examined using local examples where possible.

Towards integration of planning instruments

While ironically the purpose of the act is to achieve more integrated management of physical and natural resources, the instruments created by the Resource Management Act, including the act itself, focus primarily on the management of environmental effects. Under section 5 of the act, social and economic effects of development have less weight. Other instruments prepared under different mandates by councils, such as annual and financial plans, are focused on short (1-year) and long-term (10-year) local-authority expenditure. The only 'integrators' are strategic plans which are not mandatory, although the thinking behind these documents should form the basis of the new financial strategic plans currently being prepared by councils around the country as a consequence of an amendment to the Local Government Act in 1996. Consequently, it is likely that in time the non-mandatory strategic plans will be replaced by the financial strategic plans, but they will have served an important transition purpose for councils.

In some ways, the creation of these independent instruments can be construed as undermining the notion of integrated economic and environmental management. This is further compounded by the devolved and co-operative mandate of the Resource Management Act which sets out a relatively broad and non-

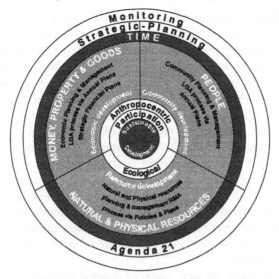

Figure 1 Resources, sustainable development, policies and plans. In this figure, resources are comprised of: (a) people; (b) money, property and goods; and (c) natural and physical resources. Sustainable development is comprised of: (a) social development (of people); (b) economic development (of money, property and goods); and (c) resource development (of natural and physical resources). Social development and economic development are anthropocentric, whereas resource development is ecocentric. Policies and plans are comprised of: (a) social planning and management through processes provided in the Local Government Act, such as social charters and annual plans; (b) economic planning and management through the Local Government Act processes via annual plans and strategic financial plans); and (c) natural and physical resources planning through the processes of the Resource Management Act. The arrows suggest integration between the three main components of the system. Central to all of the processes is public participation in decision-making and the monitoring of policies, plans and outcomes. Open to local government is the use of non-statutory tools to achieve social, economic and environmental goals, such as strategic planning, economic instruments and resource-management strategies. (Description of a diagram developed by Ernest New and Associates 1993.)

prescriptive framework within which councils develop their resource-management plans. Additionally, local government reform has devolved more functions from central to local government, while at the same time stripping away financial grants and subsidies. Thus, the onus is very much on councils to achieve the integration between their various documents.

With the pressure to meet the requirements of the Resource Management Act, many councils prepared their resource-management plans before considering the need for a strategic plan. There was no central-government directive or encouragement for them to develop strategic plans. Some councils, particularly in the metropolitan areas, with more established histories of consulting with their communities and a greater capacity to undertake this work, began with the preparation of a strategic plan. For example, Waitakere City Council in the west of metropolitan Auckland, produced a 'green print' which guided the development and co-ordination of their strategic plan, district plan and annual plan (Waitakere City Council 1996). In addition to their strategic plans, both Auckland and Manukau Cities (in the same region) use their annual plans to co-ordinate the key activities of their councils (Auckland City Council 1997; Manukau City Council 1996). In the Manukau City Annual Plan, the city projects itself as the 'face of the future' and the 'manufacturing pulse of New Zealand'. In so doing, it aims for a balance of three main activities: community development, economic development and environmental management. Auckland City wants to become the 'outstanding city of the South Pacific'. It too has identified similar objectives to achieve this goal. These initiatives are enabled by the Local Government Act, but are not mandated. Hence, not all councils would adopt these approaches.

However, by the time the next round of regional and district-plan preparation commences early next century, most councils are likely to have established processes to ensure that these plans are integrated with others. It is also likely that a significant number of these councils will have been absorbed into larger units, as a consequence of roading reforms, more unitary authorities and adjacent councils getting together, in order to increase administrative and financial efficiencies.

Implementation issues

It is unlikely that the designers of the Resource Management Act in the mid-1980s envisaged the reshaping of local government in the way it has subsequently unfolded in the 1990s. The advent of reform in local government with emphasis on cost-recovery, contracting out of services and corporatization of some activities has radically changed the focus and style of local government. This has raised major issues about local-government implementation of the new environmental mandate (OECD 1996).

A central issue focuses on the costs of administration. In a country of only 3.6 million people served by eighty-six councils, several hundred million dollars are being spent on' the preparation of the new style plans (Dixon *et al.* 1997). There is no guarantee of improved environmental outcomes, certainly in the shorter term, because it is taking years to prepare and implement the plans. Even when implemented, other contextual factors, like macro-economic forces, may have much more bearing on environmental outcomes than rules in plans. Developers too are facing much higher costs when obtaining resource consents and for on-going monitoring of them. Thus, there is on-going debate about how much money communities are prepared to invest in achieving improved environmental quality at the expense of economic development. These debates are occurring at both local and national levels.

A second issue relates to the capacity of local government for implementation of the Resource Management Act. In many ways, the drive for efficiency comes on top of increased demands on councils, such as the need to undertake increased public consultation and the monitoring of the state of the environment, consent compliance and the effectiveness of policy. Despite reforms, some councils are too small to maintain a strong professional base to implement the various new mandates (Howell *et al.* 1996). Generally, there is a shortage of highly skilled practitioners while a tight fiscal environment means that young staff are not always getting the mentoring they need (Dixon *et al.* 1997). Many councils have experienced constant restructuring so that new systems being put in place are still being consolidated. In addition, local government is proceeding rapidly with the contracting out of services, such as processing of resource consents, supply of water, engineering advice and so on. For some councils, this means a significant reduction in staff numbers. An example is Papakura City Council, a district of 40,000 people in metropolitan Auckland, that now has only twenty-two staff from more than sixty as a consequence of corporatizing most of its services.

Implementation has been exacerbated by several additional factors. These include, for example: the different, but in some instances overlapping, functions of regions and districts; professional prejudices (scientists towards planners); poor political leadership in many councils; unrealistic political deadlines imposed on plan preparation; the continued dominance of legal considerations; and inadequate resources within councils for implementation (Dixon *et al.* 1997). The pressures on local government councillors and staff to implement the new mandates and prepare plans has led to a focus on process and consultation often at the expense of substantive analysis and outcomes (Crawford *et al.* 1997). The capacity of councils to do more with less will be exacerbated further if proposed reforms for roading are implemented (Ministry of Transport 1997). These reforms could result in corporatization of funding arrangements for land transport so that councils may no longer be responsible for the management of funds for roads in their districts which, for some councils, is over 60 per cent of their budget.

A third issue is the need for professionals to reconcile paradigms of land-use planning, environmental management, and economic theory (Dixon 1997). The reconciliation of environment and development in a market-oriented context is one of the most fundamental challenges of the Resource Management Act. It may yet prove to be insurmountable because the principles of environmental sustainability and market-led development of resources are incompatible, except where the developer sees economic advantage in maintaining and/or enhancing environmental quality. The recent illegal importation of a rabbit-killing virus by high-country farmers operating in marginal environments in the South Island demonstrates the age-old problem of land users not wanting to internalize the environmental costs of their economic developments. The difficulties of implementation at a local level compound the intellectually demanding nature of the new environmental mandate.

Local tensions

What we are seeing played out in a number of localities is unheralded adverse reaction to proposed district plans. In some districts, people have been marching in the streets to protest about provisions in notified plans. Slogans such as 'can the plan' and 'legalised property theft' are emotive one-liners which capture public imagination (Newman 1997). Rural discontent over proposals to protect significant natural areas on private property are part of a strong 'anti-regulation' theme dominating national and local politics. New requirements for resource consents in rural areas are perceived by some landowners to be a form of revenue gathering by councils.

The rhetoric of the new right is distracting and simplistic (McShane 1996; Pavletich and McShane 1997), but underscores the difficulties of reconciling the twin drivers of the market-led development and sustainable management of the environment through regulatory and non-regulatory means. This dilemma is reflected in plans which need to meet the requirements of the mandate, pass the scrutiny of the legal profession and accommodate community demands for certainty. As a consequence, some plans are complex, lengthy and voluminous. While these new-style plans may be accepted by urban communities well used to controls on private property, some are meeting considerable resistance in rural areas where landowners are used to fewer regulatory controls. In addition, some of these plans carry with them high costs of implementation which are only serving to increase local tensions. Developing sophisticated plans which are expensive to administer in areas unaccustomed to a regulatory culture carries a high political risk for councils and signal the dangers of leaving communities behind when preparing these new style plans.

One of the more dramatic examples of the tension between environmental protection and economic development is provided by the reaction of farmers throughout the country to new plans which make clear the need to protect significant landscapes and indigenous vegetation. For example, in the Far North District of the North Island, farmers have: marched on town halls; invoked the support of Federated Farmers, a national farming organization; leafleted the entire district; and captured media attention, in order to make clear that they do not want to be subject to rules in plans requiring them to obtain consents for their activities. Their extreme behaviour mirrored that which occurred a year previously in the Tasman District, a council at the top of the South Island. In both instances, the plans produced by the councils are not considered by professionals to be poor ones. Rather, it suggests that they are effective instruments for helping to maintain and enhance environmental quality as directed by the act. However, reactions by some sectors of these communities suggest that the proposed means of achieving improved environmental outcomes are not acceptable. In the case of such strong opposition, rurally based councils have little option but to amend or withdraw their plans, or parts thereof.

Urban-based conflicts demonstrate the multiple roles of local councils where efforts to protect environmental values are traded off against initiatives to foster economic development. Examples include massive investments in retrofitting infrastructure in many cities throughout New Zealand as in Auckland, attempts to encroach on greenbelts as in Christchurch, and motorway extensions as in Wellington. More specifically, the proposal by the Auckland City Council to establish a downtown transport centre has drawn heated opposition from lobby groups concerned about: lack of public consultation; council involvement in economic development; adverse environmental effects; and the merits of the design and location of the proposal.

These, and other examples, highlight sharply local debate between environment and development in ways not previously observed. It is a debate that has extended to the rural areas of New Zealand. This is because the Resource Management Act is about the management of natural and physical resources, whereas its predecessor, the Town and Country Planning Act, focused more upon the built environment. The continuing tensions between environment and development illustrate that the new mandates have not fundamentally changed the arguments.

Concerns of national corporates

One of the difficulties reported by corporates involved in major resource developments, like forestry, is that they are now faced with high transaction costs. This is because they have to negotiate with each council over policies in plans as well as particular projects. Since a key rationale of the reforms was for councils to be responsible for problem-solving at a local level, it was accepted that different councils would identify for themselves different sets of issues in their plans as the basis for developing objectives, policies, methods and rules for achieving desired environmental outcomes. This approach accepted variation among different communities of interest and thereby variation among plans.

The more uniform approach guided by central government under the old planning regime provided much more certainty for corporates whose interests cut across many councils. In the new regime they are faced with greater differentiation and uncertainty. For example, some forestry companies report spending many months negotiating with councils on regional plans only to find that when they are publicly notified, the outcomes to which they had believed they were party had altered which forced them to adopt an adversarial situation.

What these corporates wish to see in plans are clear environmental bottom-lines through the specification of standards. More than that, they would like to see some uniformity of standards across councils. Interestingly, this suggests a call for national policy statements and standards which are enabled by the Resource Management Act, but strongly resisted by a government committed to market-led economic development and the devolution of decision-making to local levels. Further, the corporates, like mineral exploration and forestry companies, argue that too often their extractive industries are singled out for undue attention by councils to the advantage of other sector groups, like farmers. However, this argument is by no means universal, particularly where councils have taken seriously the need to protect landscapes and environments from traditional land-use practices like farming, as evident in the examples cited above.

Future directions

While local government is becoming more managerial and focused in its delivery of services to its communities in the context of market-led economic policy, it is too early, given a long transition period for implementation of reforms, to generalize outcomes. The discussion illustrates the diversity of approaches to environment and development which can be taken in the context of a devolved and co-operative mandate and the absence of coherent national policies. It also illustrates countervailing forces at work. On the one hand, the mandates set in place the opportunity for a wide range of variation in approaches to environmental management by local government. On the other hand, it shows the resistance to this variation which is leading to calls for review of the Resource Management Act by dissatisfied groups, as well as further reform of local government. This tension between local and national forms of governance has long been a feature of local-government reform in New Zealand. It remains to be seen whether further reforms will lead to stronger local government (Howell *et al.* 1996).

The Resource Management Act is a co-operative mandate which assumes that local councils are committed to its national goals, but do not necessarily have the capacity to comply. It also assumes that central government will provide the technical and financial resources needed for building capacity for local government to carry out its mandate. To date, many would argue that central government has fallen well short of expectations in this regard (Dormer 1994; May *et al.* 1996). The lack of support for implementation of a challenging new mandate, coupled with wide sweeping changes in local government which have not yet enhanced capacity, has made it very difficult for the majority of councils to deal with the demands of reconciling environment and development. This has become painfully apparent as new plans are publicly notified and debated.

The ascendance of the market liberals in the national coalition government in November 1997 suggests that completion of the new-right reforms is high on the agenda. Likely changes include the adoption of land transport reforms which will inevitably force cash-starved councils into amalgamations. This will invoke strong reactions from their communities concerned about the loss of local representation. The resultant restructuring through amalgamations may initially improve the capacity of councils to implement the Resource Management Act, through the creation of professionally stronger units for policy-making and regulation. However, these amalgamations will engender on-going change and uncertainty for councils and their communities. Paradoxically, if this restructuring is accompanied at the same time by further down-sizing, corporatization, and privatization of services, the capacity of local government to undertake further innovation may well be weakened. It is also possible that implementation of the Resource Management Act will come under further scrutiny by central government.

Indeed, by the time the act is fully implemented, which will take several more years, the basic concepts of the legislation will be in need of major review, if it has not already occurred earlier. Given the current set of institutional arrangements and inevitable changes ahead, it is hard to see that integrated management of environment and development will be made easier for local government in the short term.

References

Auckland City Council (1997) *This is Your City: Outstanding Auckland – Auckland City Council 1997/98 Annual Plan*, Auckland: Auckland City Council.

Bush, G. (1995) *Local Government and Politics in New Zealand*, second edition, Auckland: Auckland University Press.

Crawford, J., Dixon, J., Ericksen, N. and Berke, P. (1997) *Regional policy statements – synthesis of plan quality coding results, postal questionnaire, and interview findings*, Palmerston North: Massey University, Department of Resource and Environmental Planning; Hamilton: University of Waikato, Centre for Environmental and Resource Studies; and Chapel Hill, NC: University of North Carolina, Department of Regional and City Planning. (Material produced for PUCM I Review Workshops, unpublished.)

Dixon, J. E. 1997: 'Urban and environmental planning' in *The New Zealand Knowledge Base: Social Sciences*, compiled by R. Hill. (A report prepared for the Ministry for Research, Science and Technology. Wellington.)

Dixon, J. E., Ericksen, N. J., Crawford, J. and Berke, P. (1997) 'Planning under a co-operative mandate: New plans for New Zealand', *Journal of Environmental Planning and Management*, Special Issue, 40(5), 603 14.

Dormer, A. (1994) *The Resource Management Act 1991: The Transition and Business*, Wellington: New Zealand Business Roundtable.

Ernest New and Associates (1993) *Resources & Time, Policies and Plans, Sustainable Development*, Invercargill: Ernest New and Associates, Resource Management Consultants.

Howell, R., McDermott, P. and Forgie, V. (1996) *The unfinished reform in local government: The legacy and the prospect*, Palmerston North: Massey University, Department of Management Systems and Department of Resource and Environmental Planning. (Occasional Papers in Local Government Studies, Number 3.)

Manukau City Council (1996) *Manukau City Council Annual Plan 1996/97*, Manukau: Manukau City Council.

May, P., Burby, R. J., Ericksen, N. J., Handmer, J., Dixon, J. E., Michaels, S. and Smith, D. I. (1996) *Environmental Management and Governance: Intergovernmental Approaches to Hazards and Sustainability*, London: Routledge Press.

McShane, O. (1996) *The impact of the Resource Management Act on the 'housing and construction' components of the Consumer Price Index*, Christchurch and Auckland: Reserve Bank, Wellington. (A report prepared for the Reserve Bank of New Zealand.)

Ministry for the Environment (1994) *Taking up the Challenge of Agenda 21: A Guide for Local Government*, Wellington: Ministry for the Environment.

Ministry of Transport (1997) *Land Transport Pricing Study: Options for the Future*, Wellington: Ministry of Transport.

Newman, M. (1997) 'Resource law being used for legal theft of private property', *New Zealand Herald*, 21 July 1997.

New Zealand Business Roundtable (1995) *Local Government in New Zealand: An Overview of Economic and Financial Issues*, Wellington: New Zealand Business Roundtable.

OECD (1996) *Environmental Performance Reviews: New Zealand*, Paris: Organisation for Economic Co-operation and Development (OECD).

Pavletich, H. J. and McShane, O. (1997) *Taking the heat off the RMA*, Auckland: Pavletich Properties Ltd and McShane Venture Management Ltd. (A report for the Hon. Simon Upton, Minister for the Environment.)

Waitakere City Council (1996) *Waitakere City Council 1996/97 Annual Plan and Budget*, Waitakere: Waitakere City Council.

Part III
The UK context for integration

5 Integrating economic and environmental policy

A context for UK local and regional government

Derek Taylor

Introduction

Much of what local authorities do impacts on the environment, the local economy and the social condition of its electorate. These impacts stem from the application of council policy on aspects as diverse as:

- development plans to improve the balance of land uses throughout the council area;
- anti-poverty strategies to improve welfare and cohesion in under-privileged neighbourhoods;
- environmental health regulations to reduce pollutant emissions from factories and traffic;
- recycling targets in waste-management plans to lower the quantity of waste going to disposal;
- public open-space and landscape-area management to improve quality of life; or
- inward investment and economic-development programmes to create employment.

Such influences are obvious, intended and easy to monitor. However, local authorities exercise just as much influence through their day-to-day business activities. Impacts, here, stem from the fact that councils are probably the largest business in their area, employing several thousand people and spending millions of pounds annually on a wide range of services, materials and supplies.

What is more, policy and business processes designed to advance particular objectives have a tendency to influence other aspects, sometimes in ways that are not intended. For example:

- development plans which do not provide a mix of housing and fail to target shortfalls in particular types of accommodation risk widening social inequity;
- anti-poverty strategies which ignore the physical environment endanger the prospect of attracting inward investment to revive the economy;
- enforcing pollution controls may cause factory closures and job losses;

- restricting waste collection to inorganic material can impose unfair demands on householders, unless accompanied by measures to assist home composting and paper collection;
- over-zealous maintenance regimes creating manicured open spaces can lower wildlife diversity;
- new jobs on out-of-town, greenfield sites disadvantage members of the community without a car and take open land which may have amenity and ecological value.

The key to resolving these difficulties is a corporate approach to recognizing and integrating the linkages inherent in policy and management programmes. Until recently, councils carried out policy and business processes with little regard to this, but growing concern about the need to harmonize environmental, economic and social objectives is forcing them to develop mechanisms for responding to the overlaps.

The purpose of this chapter is to present an overview of why, and how, local authorities are addressing the practicalities of integration. The following two sections on the context and capacities for the integration of economic and environmental objectives illustrate how trends in sustainable development at international, national and community levels have influenced local government to develop new approaches. They provide a context for subsequent chapters in Part II of the book. The subsequent section which focuses on the responses to the call for integration looks at how authorities are responding to this pressure in four strategic areas, and draws some conclusions as to the effectiveness of the response. This section is a background for the case studies in Part III. The final section concludes by drawing the threads together and indicates how the response might develop in future.

Towards integration

Some 10 years ago, the world first heard about 'sustainable development', the conceptual framework for integrating environmental, economic and social policy, globally and locally (WCED 1987). The 'Brundtland' definition of sustainable development has been the starting-point for sustainability policy ever since.[1] But it was in Rio de Janeiro, in 1992, that the definition was fleshed-out. Heads of state attending the UN summit confronted the linkages between the planet's environment and the economic and social activities of its people. In terms of policy output, they succeeded. In addition to declaring its commitment to sustainable development, and issuing conventions on climate change, forests and biodiversity, the summit produced Agenda 21, its strategic blueprint for creating a more sustainable future into the next century (UNCED 1992).

Agenda 21 is a framework encouraging governments to produce national sustainable development strategies. Though targetted at national administrations, local government is called on to play a central role in delivering policy on the ground. Two-thirds of the hundreds of aspirations in Agenda 21 cannot be achieved without the co-operation of local authorities (LGMB 1993). Municipal services like planning, transportation, waste, public health, education, social welfare and economic development are acknowledged as critical implementation mechanisms. This endorsement of local environmental and socio-economic subsidiarity has been instrumental in stimulating local government to embrace sustainability as a guiding philosophy.

As the level of elected governance closest to the citizen, many councils are seizing the opportunities presented by the new philosophy to forge more productive alliances with their stakeholders. The challenge in chapter 28 of Agenda 21 that each municipality should engage with its local community to create 'Local' Agenda 21 (LA21) strategies has furnished a key mechanism for this realignment. As this chapter demonstrates, however, putting both philosophy and mechanisms into practice has not always been easy. Even so, the review of Rio in 1997 [2] revealed that while national governments are slowly getting-to-grips with their Agenda 21 responsibilities, municipalities are making better progress.

Closer to home, the European Union (EU) started to grapple with the Rio agenda before 1992. In 1987, the Single European Act placed the environment at the centre of policy development,[3] while 5 years later the Maastricht Treaty made the focus sustainable development.[4] Many of the 200+ environmental regulations and directives, stemming from the EU's integrated approach, fall ultimately to local government for implementation. Europe has also been exerting pressure on councils through its mainstream policy activity, and via the grant regimes associated with regional policy. Most profound of all has been the influence of the Fifth Environmental Action Programme (5EAP), issued in 1992. This programme and its review (CEC 1992; 1996) signals a sea-change from policies dealing with discrete environmental issues and end-of-pipe solutions towards broader measures aimed at integrated intervention to prevent problems arising in the first place. '*Towards Sustainability*', the sub-title of 5EAP, reveals the EU's determination to integrate the environment with socio-economic domains. Though aimed at member governments, like Agenda 21, a significant proportion of the measures in 5EAP will only happen through the involvement of local authorities.[5]

For local authorities, especially when operating in regional partnerships, a bonus of EU membership has been access to funds provided to implement regional policies. Though often associated in the past with infrastructure projects which were anything but environmentally friendly, the trend is shifting to support for schemes based on harmonizing the relationship between environment and economy. The requirement for environmental appraisals to accompany bids to some funding programmes is also designed to assist the process of integration. Overtly environmental funding instruments such as LIFE, and those connected with areas like the energy field, such as THERMIE, ALTENER and SAVE, can be further stimuli for local authorities to link economy with the environment.

In the UK, the first post-Brundtland manifestation of government concern for sustainability issues was the 1990 white paper on the environment (DoE 1990). Annual up-dates have tracked the progress of key measures (DoE 1996), while the first UK state-of-the-environment report appeared in 1992 (DoE 1992). However, there was no serious attempt to integrate environment with economic and social objectives in these early reports. An integrated regulatory regime for industrial discharges to air, water and land was introduced at the beginning of the 1990s, co-ordinated by a new Environment Agency. However, the benefits of 'integrated pollution control' were offset, somewhat, by transfer to the agency of the waste-regulation responsibilities of councils, thus divorcing them from waste planning and collection. The contemporary privatizations of water, energy and transport did more to retard the prospects of rapid integration in sectors critical to the achievement of sustainable development. After Rio, there was a positive shift in emphasis, though local government continued to set the pace on practical integration. In 1994, the first national sustainable development strategy appeared (DoE 1994). Though a clearer rationale of the need for integration, like the 1990 white paper, it lacked concrete proposals to bring this about.

Of more direct relevance to the efforts of local authorities to synchronize environment and economy were developments in fields like land-use planning, waste management, biodiversity and air-quality control. From the early 1990s, national planning guidance (PPGs) made sustainable development the core objective for regional planning, statutory development plans and control.[6] However, much of the early guidance was vague, leaving a lot of detail to be worked out locally. Endorsement of the 'plan-led' basis of the planning system went a long way to redressing the balance between community and market-led interests in this important area of public policy. However, a continuing ambivalence of government support for a strong local-government role in regional planning, and a preference for arms-length development corporations based on private-sector ethics, impeded progress with integration at this level. A number of regional conferences and associations produced planning guidance and strategies based on sustainable development principles, though for many the priority was economic, rather than sustainable, development.

The prospect of a more coherent national framework for integrating environmental, economic and social objectives was signalled in Labour's 1997 election manifesto and, more tangibly, in their earlier policy commission report on the environment (Labour Party 1997). Tony Blair made his first overseas prime-ministerial appearance at the UN session reviewing progress since Rio, pledging that all UK local authorities will complete their LA21 strategies by the end of 2000 (British Information Services 1997). It is still too early to judge whether New Labour is genuinely serious about integration. On some issues, progress has not met expectations. The chancellor has yet to fully confront the fiscal opportunities for driving-down resource consumption and pollution by eco-taxation, while environment ministers are struggling with the sustainability issues surrounding the location of the 4.4 million new households needed by the first decades of the next century. On the plus side stand the institution of a cross-departmental environmental audit committee, a sustainable development unit to advise ministers, a review of the 1994 sustainable development strategy, work on sustainability indicators for regular comparison with national economic indicators, production of an integrated transport policy and steps to formalize regional government. Important changes in emphasis have also emerged in departments once hostile, or neutral, to the call for sustainable development. But while the Foreign Office and the Department for International Development are moving forward, critical players like the Treasury and Department for Trade and Industry (DTI) remain agnostic.

But the most significant role that government has played over the past decade has been behind the scenes. Co-operation between the Department of Environment, Transport and Regions (DETR) and local government on how local authorities should address sustainability issues has been remarkably good, at a time when general relationships have been hostile. This has been due to the strong, politically neutral, lead taken on sustainability by the local-authority associations (now combined into the single Local Government Association) and COSLA, expertly advised by the Local Government Management Board (LGMB). Both sides have co-ordinated action through a joint forum, which has overseen a range of guidance and advice on sustainable development issues.[7]

At the local level, international protocols, EU interventions and UK policy are not the only pressures influencing local authorities to integrate their environmental, economic and social activities. The very nature of local councils makes them pivotal to the sustainable development aspirations of their local communities. Paradoxically, while engagement in the local democratic process is poor, the degree of public interest in the environment has grown since the 1980s. While it is entirely healthy for councils to have to respond to a more environmentally aware electorate, and a growing number of more articulate pressure groups pushing specific agendas, councils have not been particularly good at persuading people of the synergy between the environment, economic activity and social equity. Responsibility for redressing this, of course, has to be a priority for local government, itself, and LA21 is a key technique for doing so. But until councils become more responsive organizations, and operate in more openly democratic ways, they will struggle to make the message stick. The other prerequisite, if integration is to succeed, is the forging of real partnerships and understanding between councils and local businesses. Though examples of good practice stand out, they do so because they are the exception, rather than the rule.

The capacity for integration

Setting aside the imperative of having to react to so much external pressure, particularly where it has been made a legal requirement, many councils have realized that perpetuating the *status quo* is no longer an option. If we are to reverse the dangerously unsustainable trends which are currently at work, local leadership and stewardship are essential. The more visionary councils have recognized that there is also opportunity and self-interest in tackling the issues. They have seen that sustainable development offers a proactive route to engagement with stakeholders and the public, and for improving the culture and effectiveness of their organization at the same time. 'Putting your own house in order' can do more than get campaign groups off a council's back, or improve its image. It can create genuine credibility, enabling the authority to cajole local business and the public to begin to make the more sustainable choices that lie at the root of achieving the vision in Agenda 21 (Taylor and Lusser 1998).

As large businesses, councils need to obtain, and provide, a wide range of best value services. Sustainable development has to be based on seeking a continuous improvement in performance, making today's targets tomorrow's baselines. This new dynamic enhances the position of the successful authority in the market place, while basing service delivery on the integration of environmental and socio-economic objectives can generate a more innovative internal culture of ideas and solutions. The Audit Commission argues that local authorities spend a consider-able part of their budgets on energy, water, waste disposal and in-house transport (Audit Commission 1997). Introducing efficiencies in these four areas alone will not only save money; it will go a long way towards reducing a council's considerable ecological footprint. In-house skills gained from integrating policy and action on the environment with social and economic dimensions will enrich the knowledge base of the organization and motivate staff to new ways of working and collaboration. Some of these skills are transferable, providing potential for income generation. Integrating these aspects is now also a requirement of some funding, so operating in this way improves the chance of succeeding with bids in an increasingly competitive area.

The capacity of local authorities to respond to the challenge of sustainable development has been facilitated by their being responsible already for many aspects of public policy that need to be integrated if the challenge is to be met. Environmental and public-health duties go back to the origins of modern municipal government, while the housing, social welfare, education, land-use planning, economic development, transportation, conservation and amenity functions which have been added more recently complete a powerful array of opportunities for intervention. Set against this, however, has been the stripping away, since 1980, of many town-hall powers by an executive intent on favouring the private over the public sector. A further barrier was erected by the squeezing of resources through capping and rate-support-grant reductions, which bit hardest when new legislation and guidance were encouraging councils to pursue integration. Responding to the encouragement was not helped by the fact that government departments, preaching the need for integration, were not practising their own message.

That councils have been trying to integrate sustainability into their policy and practice suggests that more cash and unequivocal support from the centre are not essential to progress. Having the right attitude, and the will to go for integration, have been more important drivers. Rearranging budget priorities to reflect a new philosophy is not an insurmountable problem. For many, adopting sustainable development has not been so much a matter of choice as a necessity given that the imbalance between environmental, social and economic conditions has moved so far out of line. However, the councils who are succeeding best are the ones who have adopted sustainability as a corporate value. Those who rely on part of the organization to make the running invariably struggle with integration.

A further barrier has been the hierarchical structures which have traditionally characterized local authorities, engendering in their turn the formation of independent groupings of specialized technical and professional expertise. Integration requires negotiation, collaboration and inter-disciplinary consensus, which hierarchies tend not to encourage. It demands new ways of thinking, and a vision that transcends specialisms, deploying different strengths to achieve a common purpose. The vision has also to extend beyond politicians and officers to involve the local community. These are constituencies with which local authorities lost touch in the 1970s and 1980s, as energies were harnessed to deal with attacks from central government. The new integrating philosophy of sustainability will not work without strenuous efforts at re-engagement, and the creation of cultures in which more responsive ways of working can flourish.

This need for greater openness in policy development, inside and outside authorities, is leading to experimentation with a range of innovative structural mechanisms and the realignment of existing ones. Research shows how serious the alienation of local people from municipal democracy has become. In some ways, the very *raison d'être* of local government, that it exists to provide services which are needed and valued by local people, is threatened by a ground-swell of mistrust on the part of those who are being 'served' (Grove-White 1997) and many have lost faith in the system and doubt its willingness and ability to represent their interests at all (MacNaghten *et al.* 1995). So, in prompting councillors and officers to re-think how they interface with, and serve, their constituencies, sustainable development is becoming a way of helping to address the democratic deficit which has grown-up over recent decades. New Labour's decision to move quickly on local-democracy issues, through devolution, regional governance, elected mayors, reviewing council financing and the 'Best Value' concept (DETR 1998), will add impetus to grass-roots efforts in this important area.

What are these emerging structures which are helping to create the framework for promoting sustainable development? They range from re-engineering the whole authority to single measures like employing a specialist officer to co-ordinate sustainability activity. At the macro-level, reforming monolithic, hierarchical, departmental structures into multi-purpose, multi-disciplinary teams, coupled with business-planning methods and target-based performance regimes, is now widespread. Integrating committee structures by amalgamating services, or creating an over-arching members' group for co-ordinating sustainable development, is also becoming common. A few councils have broken free from traditional decision-making regimes by moving to a streamlined, executive system where members and senior officers operate like private-sector boards. Increasingly, councils are recasting corporate mission statements and business objectives to make sustainable development the principal goal. Use of some of the more influential sustainable development planning and delivery mechanisms is considered in the next section.

As far as external relationships are concerned, a growing tendency has been the emergence of more vigorous partnerships with business and community groups in the form of business clubs, public forums and round tables. Focus groups, visioning conferences, citizens' juries, town meetings, 'planning for real' exercises and community planning techniques are all being used increasingly to engage stakeholders in deciding which services should be provided and how. Some councils have devolved service delivery to largely self-sufficient neighbourhood offices. At the regional level, groupings of local authorities, private sector and voluntary interests are beginning to work together on strategic frameworks for integrating economic regeneration with environmental protection and social renewal.

Strategic responses to integration

Reference has been made to how LA21 strategies can forge new stakeholder partnerships for integrating environmental, economic and social policy. Though LA21 emerged only 5 years ago, and there is no statutory duty to produce a strategy, nearly three-quarters of UK councils are doing so (LGMB 1997). At the heart of LA21 is a collaborative process between councils and communities to agree a long-term vision and action plan for sustainable development. The key to how this will help to integrate policy lies in the objectives which are set for the strategy, and the composition of the partnership driving it. Many are working with the thirteen goals for creating a sustainable community established as part of a national project (LGMB 1995). Three of these goals cover the environmental domain, three the economic, four concern social equity and the final three deal with quality of life. The difficulty arises in terms of how partnerships are able to integrate the different goals and convert them into achievable policy.

This raises a number of issues around the practice of LA21 which need to be addressed if it is to realize its full potential as an engine for change. Irrespective of the national guidance which clearly states that LA21 must deal with all the components of sustainable development, some councils over-concentrate on the environmental dimension. Maybe this is not surprising given that many councils allocate responsibility for LA21 to the people who dealt previously with the environmental portfolio. It is also an area over which many councils feel it is easier to exercise influence. Long-held specialisms, practised in compartmentalized, hierarchical structures, compound the problem of constructing the holistic partnerships essential to proper integration. Crucial connections with the planning service and economic development functions have proved difficult to forge in many authorities.

LA21 covers an enormous canvas, and unless there is consensus on clear priorities from the outset, expectations can quickly flounder in a wealth of debate and little direct action. Some LA21s have tended to concentrate on incremental, small-scale action, at the expense of confronting strategic linkages. The technique was officially re-launched at the start of 1998 (DETR, LGA and LGMB 1998), partly with the aim of forcing activity more onto the social and economic parts of the agenda. The fact that the majority of those at the re-launch were from the corporate arena – council leaders, committee chairs, chief executives and senior officers (Levett 1998), together with the prime minister's call for total compliance, should also help to raise its profile. These are encouraging signs, because there is no doubt that LA21 has the potential to become the most powerful integrating force available to a local authority and its community for achieving sustainable development.

For many years, the environmental management role of local government was confined to protecting and conserving wildlife habitats, open space, landscapes, tree cover, listed buildings and archaeological sites. Countryside management strategies and implementation programmes are now well-established as strategic responses to the conservation of biodiversity, heritage and local amenity. They are critical to any coherent sustainability strategy. However, now the agenda has expanded to encompass linkages between the environment and other themes, councils have had to extend their response. In particular, if a local authority wishes to control its own influence over sustainability, it must review its own environmental impacts and institute programmes to improve its performance.

To do this, councils are increasingly adopting a systematic approach based on international environmental management systems like ISO 14001, or the EU eco-management and audit scheme (EMAS). The latter has been customized for use by local government (Department of the Environment and Local Government Management Board 1993), and provides the most commonly used basis. EMAS does not deal overtly with the performance of council policies and practices in the socio-economic arena, though practitioners are beginning to think about how it might be extended to do so (Levett 1996). Even with its bias towards the environment, however, EMAS can reveal significant overlaps, contributing to a council's attempts to integrate its wider policies. The most obvious way to do this is to apply environmental management techniques to the council's economic planning and development functions. This will immediately highlight opportunities for harmonizing environmental and economic goals, providing a basis for the more integrated policies needed to implement them.

There are more subtle ways in which EMAS can help. Buying-in large quantities of energy, materials, resources and other supplies makes a major contribution to a council's environmental footprint. Because local authorities are large businesses, and purchasers, they exercise considerable economic clout. Some, like the London Borough of Sutton, turn this clout to environmental advantage by re-writing contract specifications to require their suppliers to operate in a more environmentally friendly manner. The potential for exerting pressure on the local economy to green-up its act is considerable, though it can only be done by a council which has bitten the bullet on environmental management itself. A more profound way to internalize the integration between environmental, social and economic policy would be to subject all council budgetary processes to sustainable development appraisal. Greening every budget line, and the council's huge investment portfolio, should be essential for any council wishing to be serious about its sustainability principles. No council has yet gone this far, though some, like Nottinghamshire County Council, are experimenting at the margins.

Councils also need to pursue integration in the important area of appraising and monitoring the local environment. Environmental appraisal of major proposals, and development plans, is now required by law. Though not designed specifically as a tool for integration, appraisal techniques can highlight, and facilitate consideration of, the relationship between the environmental and socio-economic advantages of a proposal. A further technique which focuses more sharply on this relationship is the definition of indicators to monitor progress on local sustainability. Some councils have started to monitor and publish these in local sustainability reports (LGMB 1996), of which Lancashire County Council's is the most comprehensive (Lancashire County Council 1996). Here, an attempt has been made to evaluate the linkages between different indicators, revealing powerful correlations across a range of environmental, social and economic trends (Mullaney 1997). For example, wards experiencing high levels of unemployment also have the largest rates of low-birth weight, premature deaths, burglary and alienation from the democratic process. These 'hot-spots' of poor indicator performance are helping Lancashire, and its stakeholder forum, to re-define priorities for public/private-sector intervention in these localities; a practical demonstration of how one attempt at integration is influencing sustainable development policy in a fundamental way.

In the 50 years since its introduction, the town and country planning system has been the principal tool of public policy for directing land use to achieve environmental and socio-economic goals. Development planning and control have been used by central and local government to provide sites for housing, jobs, transport and other social and economic infrastructure, while protecting the best landscape, townscape and open space. The 'plan-led' system, especially since sustainable development became its core principle, is now central to the integration of environmental, social and economic aspirations. Greater emphasis on mixed land use, locating development where it reduces the need to travel and encourages public transport, reversing the trend to out-of-town development and re-invigorating urban centres will all help to improve sustainability. Planning has so much to offer because it contributes to each of the four themes of sustainable development, in that it:

- gives precedence to environmental protection;
- seeks to extend social equity through influencing housing provision, land for jobs and the availability of economic and welfare services;
- tries to enhance quality of life by improving general amenity; and
- has an in-built disposition towards futurity, by virtue of being a long-term, regularly up-dated, process.

However, a number of systemic constraints prevent planning from making a more significant contribution to integrating environmental and socio-economic poli-cies. The most serious is the fact that though predicated on the delivery of sustainable development, development plans are restricted to allocating uses for land. They cannot propose policies for how land is managed or influence how and why people and goods move between sites. So they are unable to deal with many of the key linkages that need unravelling if sustainability is to be achieved. Though getting land uses right is critical to reversing unsustainable trends, it is only part of the solution. Those who try to use the planning system to pursue economic and social issues essential to the creation of sustainable communities, find that it always gives precedence to the physical environment (Waterman 1997). Until development plans are able to look more holistically at these connections, their potential for creating the strategic sustainable development frameworks demanded by PPGs, will remain unfulfilled (Baker 1997). Formalizing a working interface between LA21 and the planning system would be a particularly helpful step. At the same time, improving the horizontal integration of the different functions within local-authority planning departments, and between these and other skills like environmental health, building control and economic development, would help land-use planning make a more coherent contribution to the pursuit of sustainable development (Knowland and Therivel 1998).

A further drawback is the continuing lack of a regional planning framework with bite. Resolving the most intractable sustainable development-planning issue of the moment, such as finding sites for the extra households needed over the next two decades, would be easier if there was an open and accountable mechanism able to determine housing allocations as part of a regional sustainability strategy, with powers to impose them (EDAW 1997). In this respect, the June 1997 DETR consultation paper on regional development agencies went some way towards proposing enhanced status for regional planning. However, it fell disappointingly short of recommending the production of over-arching strategic plans (Roberts 1997). The dilemma over the household projections raises in sharp relief a final big challenge facing planning practice at the moment. All parts of the system need to internalize sustainable development into everything they do, but at present this is overwhelmingly the domain of policy-makers. In due course, their efforts will trickle down to the designers, implementers, analysts, land managers and regulators, but the integration would be richer, more powerful and would happen sooner if all disciplines were working together from the outset. Some notable attempts have been made to rationalize sustainability into non-policy fields, but these are the exceptions to the rule (Barton *et al.* 1995).

Local businesses have a major impact on the environment of their area and beyond. Their collective use of energy, transport, materials and the solid, liquid and atmospheric wastes they create, contribute significantly to the environmental footprint of their locality. As society will not survive without the economic activity they generate, it is clear that businesses are central to the delivery of sustainability. It follows that if local authorities are to succeed in integrating the environment with socio-economic objectives, they must forge cohesive partnerships with their business communities. They must also deploy the policy mechanisms at their disposal to implement the results of these partnerships. LA21 strategies are one of the best frameworks for developing collaboration and action programmes.[9] Land-use plans play their part by helping to generate more sustainable economic spatial patterns, facilitating the communal infrastructure and environment in which business can flourish. Councils also produce annual economic-development plans, in consultation with business, which offer another direct opportunity for harmonization. At a more practical level, local authorities bid for resources to implement economic-development programmes from sources like the EU structural funds. Success depends on the proposals being consistent with the EU's commitment to sustainable development.

Defining a common base for co-ordinating the policies and aspirations of these different instruments is the key to maximizing their potential for integration. Surveys have found, however, that less than three-quarters of statutory develop-ment plans and under a half of economic-development plans contain integrative policies (Gibbs *et al.* 1995). Only 45 per cent of local authorities have instituted formal partnerships with business in their LA21 strategies (Tuxworth 1996), while a study of seventy EU funding bids reveals that integration is not being fully achieved here, either (Rumble and Jackson 1997). This is mainly due to uncertainty about what sustainable economic development actually is, and a reluctance to try new approaches to traditional problems. If the goal of integration is to be met, then local-authority economic-development mechanisms need to confront these issues head-on. One of the outcomes of the structural fund study is the suggestion that there is a need for an agreed definition of what actually constitutes sustainable economic development,[9] which would seem to be a minimum requirement for achieving integration. Another key principle that would accelerate the process is the importance of being able to identify, and understand, where the economic aspects of a particular activity coincide with their social and environmental counterparts. Formulating economic plans and pro-grammes on this principle will guarantee that when their economic policies and actions are implemented, environmental and social objectives will be met at the same time.

As far as translating the goal of integration into policy is concerned, there are a number of key aims that can provide a cohesive approach to environment and the economy (LGMB 1994). These are:

- helping local businesses to reduce their environmental impacts;
- encouraging a more sustainable mix of businesses in the local-authority area;
- promoting the growth of environmental industries;
- protecting the environment in ways that do not threaten jobs;
- developing business opportunities from environmental protection and enhancement.

Procurement activities are another way of greening local business as part of the strategic approach to environmental management mentioned. Considerable potential exists to use the spending power of local authorities, especially through consortia, to boost the market for environmental technology by switching to low-energy/high-efficiency equipment, recycled materials and benign goods and services. There are opportunities to develop innovative economic instruments like regulations, licences, permits and incentives to engineer a switch to more sustainable actions and choices (Roseland 1996). This is starting to happen in a small way by, for example, raising parking charges to encourage people out of cars and onto public transport. More radical, though, is the scope for using community funds to encourage a shift in favour of locally controlled and sourced enterprises. This is more common in Canada and the United States (Mathewson and McGonigle 1997), but is starting to gain ground here through the direct support some councils are giving to local enterprise and trading schemes (LETS) (Lang 1994), farmers' markets and community-food production/distribution projects.

Conclusion

The efforts of local authorities to respond to sustainable development are beginning to impact on their policies and programmes. The stimuli come from a variety of sources, operating at global to local levels. Frameworks from the Rio process, and the EU, recognize the significance of local action and provide a strategic rationale. However, dealing with intractably unsustainable global trends is a matter for the international community, and while world governments fail their Rio commitments, municipal governments face an uphill struggle. In the UK, the turbulent central–local-government relationships of recent decades produced both pressure to integrate, and diminishing powers and resources with which to respond. This confused message was not helped by the fact that calls on local government to integrate were not reciprocated at the centre, while practical support lacked coherence.

But the most important players in local sustainability are councils themselves, and the people they represent. There is now widespread support for the concept of integration, though 'jobs, not butterflies' can still be the slogan when it comes to choosing between economic development and environment. Operational progress in integrating environmental and socio-economic policy is patchy, which is perhaps not surprising, given that:

- the concepts underpinning integration are new, complex to internalize and not yet fully understood in practice;
- adjusting entrenched procedures to assimilate the concepts is not straightforward and long-established council structures are not geared-up to operating synergistically;
- like most organizations, local authorities inherently resist change;
- politicians and officials are at the bottom of a steep learning curve, and antipathy to new ideas can be strong where they may be seen to threaten well-established specialisms and interests;
- though integration is legally required in some areas, it remains largely a voluntary matter;

- local government is not held currently in sufficiently high regard by its stakeholders, whose co-operation is essential to successful integration.

More positively, the track-record of councils in LA21 and environmental management is beginning to weaken these obstacles. Progress in planning and economic development is accelerating also, though the opportunities for interface between the two, and with LA21 and EMAS, is not keeping pace. However, there are developments on the horizon which will combine to speed-up integration, including:

- internationally, the continuing work on Agenda 21 by the UN Commission on Sustainable Development coupled, hopefully, with more effort at integration by world governments[10] – the shift towards eco-taxes, moving integration closer to the heart of global financial decision-making – the spread of accessible environmental technology and eco-efficiency practices like those which demonstrate a four- to ten-fold reduction in the environmental impact of economic activity (von Weizsacker *et al.* 1997), using fewer finite resources;
- in Europe, the continuing harmonization of policy, regulatory and financial instruments around the themes of sustainable development, extending the strategic framework for integration and subsidiarity and increasing its implementation by member states;
- nationally, increasing government sympathy for integration, clarifying the strategic context – wider democratic renewal regionally and locally, encouraging more productive partnerships for integration – the application of 'best value' principles, making council services more in tune with the environment and the socio-economic needs of their electorates; and
- locally, businesses switching to greener production and management as customers, regulations and the fiscal system dictate, making it easier for local authorities to collaborate on integration – growing public awareness leading to more sustainable life-style choices, spreading as the social responsibility component of the school curriculum begins to influence behaviour and attitudes.

Whatever happens, one thing is clear. The tentative context which currently guides local authorities to pursue the integration demanded by sustainable development will strengthen, making the contribution of councils to the process more, not less, influential.

Endnotes

1 'Development which meets the needs of the present without compromising the ability of future generations to meet their own needs'. Subsequent fine-tuning has led to a consensus around a definition of sustainable development which encompasses the four themes of environment, equity, quality of life and futurity.

2 Considered by the UN General Assembly Special Session, meeting in New York in June 1997.
3 Articles 130 r, s and t introduced a new environmental title, underpinning all policy and legislation.
4 Article 2 of which introduced the goal of 'sustainable and non-inflationary growth respecting the environment' as a consideration in EU policy development.
5 It has been estimated that 40 per cent of the 5EAP can only be implemented by local authorities, either on their own or with others. See: Local Government Management Board (1993) *Towards Sustainability. The EC's Fifth Action Programme on the Environment: A Guide for Local Authorities*, Luton: LGMB.
6 This applied to all Planning Policy Guidance notes (PPGs) as they were reviewed during the 1990s. The main PPGs in question were PPG1 (General Policy and Principles), PPG6 (Town Centres and Regional Development), PPG9 (Nature Conservation), PPG12 (Development Plans and Regional Planning Guidance), PPG13 (Transport), PPG22 (Renewable Energy) and PPG23 (Planning and Pollution Control).
7 Such as *A Framework for Local Sustainability* (1994) and *Local Agenda 21 Principles and Process. A Step by Step Guide* (1994) both LGMB, Luton; and the other LGMB publications referred to throughout this chapter.
8 Good examples of proactive liaison are the Business and Environment Support Team operated by Bradford MBC, and the independent company facilitated by LB of Sutton to train businesses in environmental management.
9 Rumble and Jackson (1997) define it as 'Economic development which ... requires a pattern of development which improves the quality of life of people in all countries and regions of the Community through the creation of high and permanent levels of employment by non-inflationary economically-viable activity which:

- conserves non-renewable resources and maximizes the benefits obtained from the efficient use of all resources through the reduction of waste;
- protects highly valued environmental resources and improves the overall stock of environmental resources by minimizing pollution and emissions, and by enhancing degraded and damaged environments; while
- recognizing the rights of all people to satisfy their everyday needs for health, education, housing, food and community and spiritual well-being, both today and in the future'.

10 The OECD held a workshop in Sweden recently to begin the process of promoting the application of environmental-management techniques by its twenty-nine member governments and their agencies, which are the most economically powerful and environmentally influential on earth.

References

Audit Commission (1997) *It's a Small World. Local Government's Role as a Steward of the Environment*, London: Audit Commission.
Baker, M. (1997) 'Development plans – An agenda for future change', *TCPA Journal*, 66(9), 242–4.
Barton, H., Davies, G. and Guise, R. (1995) *'Sustainable Settlements: A Guide for Planners*, Luton: LGMB/University of the West of England.
BIS (British Information Services) (1997) 'Text of Speech by the Prime Minister, the Right Hon. Tony Blair MP, to the UN Special Session on the Environment (UNGASS), Monday 23 June 1997', New York: BIS.

CEC (Commission of the European Communities) (1992) 'Towards sustainability: An EC programme of policy and action in relation to the environment and sustainable development', Brussels: CEC.

CEC (Commission of the European Communities) (1996) 'Progress report from the Commission on the implementation of ... towards sustainability', Brussels: CEC.

DoE (Department of the Environment) (1994) *Sustainable Development: The UK Strategy*, London: HMSO.

DoE (Department of the Environment) (1992) *The UK Environment*, London: HMSO.

DoE (Department of the Environment) (1996) *This Common Inheritance: 1996 UK Annual Report*, London: HMSO.

DoE (Department of the Environment) (1990) *This Common Inheritance*, London: HMSO.

DoE/LGMB (Department of the Environment and Local Government Management Board) (1993) *A Guide to the Eco-Management and Audit Scheme for UK Local Government*, London: HMSO.

DETR, LGA and LGMB (1998) *Sustainable Local Communities for the 21st Century – Why and How to Prepare an Effective LA21 Strategy*, London: LGMB.

EDAW, Global to Local and de Montfort University (1997) *Living Places. Sustainable Homes, Sustainable Communities*, London: National Housing Forum.

Gibbs, D., Braithwaite, C. and Longhurst, J. (1995) 'Integrating economic development and environment: Local authority responses in England and Wales', paper given at first meeting of the Sustainable Cities Network, Manchester Metropolitan University.

Grove-White, R. (1997) 'Currents of cultural change', *TCPA Journal* 66(6), 169–71.

Knowland, T. and Therivel, R. (1998) 'A bit of writhing among the cables', *TCPA Journal* 67(2), 54–5.

Labour Party (1997) *Because Britain Deserves Better*, London: Labour Party.

Lancashire County Council (1996) *Lancashire's Green Audit 2: A Sustainability Report*, Preston: LCC.

Lang, P. (1994) *LETS Work. Rebuilding the Local Economy*, Bristol: Grover Books.

Levett, R. (1996) 'From eco-management and audit (EMAS) to sustainability management and audit (SMAS)', *Local Environment*, 1(3), 329–34.

Levett, R. (1998) 'LA21 passes the smooth grey suit test', *TCPA Journal*, 67(2), 52–3.

LGMB (1993) *Agenda 21 A Guide for Local Authorities in the UK*, LGMB, Luton: Local Government Management Board.

LGMB (1994) *Local Agenda 21 Roundtable Guidance: No. 3 Greening the Local Economy*, Luton: Local Government Management Board.

LGMB (1995) *Indicators for LA21 – A Summary*, Luton: Local Government Management Board.

LGMB (1996) *Sustainability Reporting. A Practical Guide for UK Local Authorities*, Luton: Local Government Management Board.

LGMB (1997) *LA21 in the UK. The First 5 Years*, London: Local Government Management Board.

MacNaghten, P. *et al.* (1995) *Public Perceptions and Sustainability in Lancashire*, Preston: Lancashire County Council.

Mathewson, A. and McGonigle, M. (1997) 'Eco-investing: Financing sustainable economic development', *Local Environment*, 2(2), 155–70.

Mullaney, A. (1997) 'Auditing equity', *TCPA Journal*, 66(6), 162–3.

Roberts, P. (1997) 'Welcome first steps', *TCPA Journal*, 66(9), 224–6.

Roseland, M. (1996) 'Economic instruments for sustainable community development, *Local Environment*, 1(2), June 1996, 197–210.

Rumble, J. and Jackson, P. (1997) 'Greening the structural funds', *TCPA Journal*, 66(9), 245–6.

Taylor, D. and Lusser, H. (1998) *Corporate Approaches to Local Agenda 21 through the Implementation of EMAS*, London: LGMB.

Tuxworth, B. (1996) 'From environment to sustainability: Surveys and analysis of LA21 process development in UK local authorities', *Local Environment*, 1(3) 277–97.

UNCED (1992) *Agenda 21 – Action Plan for the Next Century*, New York: United Nations Commission on the Environment and Development.

von Weizsacker, E., Lovins, A. B. and Lovins, L. H. (1997) *Factor Four: Doubling Wealth, Halving Resource Use*, London: Earthscan.

Waterman, P. (1997) 'More people for living communities', *TCPA Journal*, 66(6), 172–6.

World Commission on Environment and Development (1987) *Our Common Future*, Oxford: Oxford University Press.

6 Integrating environment into local economic-development policies

Experience from UK local government

David Gibbs and James Longhurst

Introduction

In the 1990s the environment came to the forefront of policy-makers' consciousness, given a powerful boost by the 1992 Earth Summit conference in Rio de Janeiro and the subsequent widespread adoption of sustainable development as a guiding principle for policy development. The concept of sustainable development became particularly attractive to policy-makers as its adoption seemingly allowed the reconciliation of environmental protection with economic development. While subsequently defining sustainable development in operational terms proved problematic, it was rapidly embraced by international, national and local policy-makers. Such policy measures frequently saw the local scale as the most appropriate for the delivery of sustainable development policies and initiatives, with a particular stress upon local authorities as the major contributor to this process (LGMB 1993). For example, implementing the Agenda 21 principles agreed at the Earth Summit requires active local-scale participation, notably through the Local Agenda 21 process (UNCED 1992). The European Union's Fifth Environmental Action Programme *Towards Sustainability* stresses the role of local authorities in integrating economic development with environmental protection (CEC 1992). In the United Kingdom the government white paper, *This Common Inheritance*, the UK sustainable development strategy and planning guidance notes all emphasize the need for integrative action at the local scale (HM Government 1994).

The sources of pressure for integrative activity have not come solely from such 'top-down' initiatives. In the UK in particular, local authorities have responded to these policy developments by implementing their own environmental strategies and initiatives (Agyeman and Evans 1994). Many of these have revolved around traditional environmental issues; for example, open space and physical planning issues (Healey and Shaw 1994). Indeed, the take-up of sustainable development by planning departments in the UK has frequently reflected a desire to relegitimate local planning in the face of greater centralization by the former Conservative government, rather than heightened concern for the environment.

However, despite the role given to local government through higher-level policies and developed through their own strategies, little is known about the capacity of local government to implement integrative policies, or the nature of their response, in any systematic manner. The ability to integrate the environment and economic development at the local scale is an implicit assumption within international and national policy. In reality, little is known about the structures and practices within local government which may act to facilitate or hinder such integrative activity. The research on which this chapter is based provides evidence of the extent to which integration is taking place within urban local authorities and examines the merits and shortcomings of integrative approaches. Our intention is to examine the nature and form of the response that has developed to date in the UK and to indicate some of the prospects for future progress at the local level. Such an evaluation is timely given that the initial enthusiasm for environmental issues generated by the Local Agenda 21 process in the UK appears to be waning in local government (Agyeman and Evans 1997).

The policy context for local authorities and their response

Within the UK, a major source of pressure for integrative activity has come through the policy process emphasizing the local scale as the most relevant for action. While the most appropriate spatial scale for implementing sustainable development is a more contentious issue than this suggests, it is certainly correct that an important component of policy design and implementation will occur at the local level (Gibbs *et al.* 1996). The Local Agenda 21 process, engendered by the Earth Summit, by definition expects a strong lead from local authorities, while the EU's Fifth Environmental Action Programme (EAP), entitled *Towards Sustainability*, identifies a number of key areas where local authorities have an important role, such as spatial planning, economic development, infrastructure development, industrial pollution, waste management and transport (CEC 1992). The European Commission estimates that implementing approximately 40 per cent of the Fifth EAP will be the responsibility of local authorities (Hams and Morphet 1994). In the United Kingdom, the former Conservative government stressed the importance of local government in implementing sustainable development in policy documentation from the white paper, *This Common Inheritance*, onwards (HM Government 1990; 1994). A cynical interpretation of this would be that central government sought to pass the responsibility to lower tiers of government without providing the capacity to respond. More recently, the new Labour government has brought the environment and regional development together under a single ministry, the Department of Environment, Transport and the Regions (DETR), suggesting that environmental issues will continue to have a focus at a sub-national scale.

An additional source of pressure for integrative activity in the UK has come from the Local Government Management Board (LGMB), both in its own right as a co-ordinating body promoting UN and EU environmental policy and good practice (LGMB 1993), and through the LGMB's work in co-ordinating the UK local-authority associations' Local Agenda 21 Steering Group (Gibbs and Healey 1995). For example, the LGMB has produced guidance for local-authority officers on 'greening economic development' which provides suggestions and ideas for integrative policies and initiatives. This work identifies five possible aims which can act as a basis for an integrated approach to environmental and economic strategies: helping businesses to reduce their impacts on the environment; encouraging a move towards a more sustainable mix of businesses in the area, for example through inward investment; fostering the development of environmental industries in the local area; protecting the environment in ways that do not threaten jobs; and seeking business opportunities through environmental protection and enhancement (LGMB 1993).

A third source of pressure for integrative activity has come from within local authorities themselves, engendered by varying mixes of elected representatives, professional officers and local pressure groups. Many local authorities have devised and implemented their own strategies to take greater account of the environment within their own activities. From the 1980s onwards, local authorities increasingly produced environmental statements and strategies, although with varying degrees of commitment to subsequent action. However, an assessment of published and internal local-authority environmental statements and strategies concluded that few of them explicitly addressed economic-development issues (Gibbs 1993). Environmental strategies and economic-development plans appeared to be generated from different perspectives – a reflection of their historic division and roots. The former largely evolved from environmental departments with an emphasis, often derived from their environmental health function, on protection. From the perspective of environment departments, business has historically been seen as a problem, whose negative impacts need to be controlled and curbed. From an economic-development perspective, business is a source of wealth creation and employment generation which needs to be promoted and encouraged.

An analysis of the policy background to local-authority activity therefore indicates a strong presumption that strategies and initiatives to integrate economic development and the environment can, and will, occur through the actions of local government. However, there is little evidence available of how, and where, initiatives which integrate the two are being introduced, or indeed whether they are being produced at all. Similarly, these policy initiatives assume that local authorities have the capacity to respond. The next section of this chapter provides empirical detail of the nature of the response by UK local authorities.

Investigating the relationship between local economic development and the environment in UK local authorities

The research on which this chapter is based was undertaken as part of the Economic and Social Research Council's Global Environmental Change Programme. The research project had two main aims: to investigate the relationship between environmental and economic strategies within urban local

Table 2 Postal-survey response rate by local-authority type

Local-authority type	Absolute number	Response rate	
		Number	%
London boroughs	33	20	60.6
Metropolitan boroughs	36	20	55.6
Non-metropolitan cities	27	20	74.1
Total	96	60	62.5

Source: Postal-survey data.

authorities in the UK; and to suggest possible frameworks for the implementation of sustainable city strategies. The methodology adopted had three component parts. First, an interview survey of fourteen local authority representative bodies was conducted to provide background detail on local-authority environmental policy and practice, to identify the sources of pressure for local government and to provide some indication of the range of responses within individual local authorities.[1] This survey confirmed the major role within the UK being played by the Local Government Management Board, both through promoting UN and EU environmental initiatives and through undertaking the work of the inter-organizational Local Agenda 21 Steering Group (LGMB 1993; Gibbs and Healey 1995). Contact with these fourteen representative bodies assisted in developing the second stage of the methodology, the postal survey. A postal survey was sent to a total of ninety-six urban local authorities in England and Wales.[2] The survey was sent to environmental, economic development and planning departments within each local authority with the aim of obtaining a collective response across all three departments. Amongst other issues the survey asked for: detail of progress in integrating environmental aims and objectives into economic development policies; current initiatives integrating economic development and the environment; and structures within local authorities for implementing sustainable development policies (Gibbs *et al.* 1996). A total of sixty completed questionnaires were returned (Table 2).

One aim of the survey was to identify 'most developed' or 'best practice' scenarios for further case-study investigation. In reality it proved difficult to identify 'best practice' given the limited extent of integrative activity in evidence. Case-study local authorities were chosen on the basis that they possessed: policies and strategies to integrate economic development and the environment; both local authority and externally led integrative initiatives in place; and an internal organizational structure which aimed to integrate the two areas. Only ten to twelve local authorities met these criteria and, while we attempted to include a range of spatial location and type of local economy, we were heavily dependent upon this limited number of local authorities agreeing to become case studies. The final element of the methodology involved a number of in-depth semi-structured interviews with six local authorities.[3] Between two and four interviews were conducted in each case study with environmental co-ordinators, planners and economic-development officers. Obviously, both the respondents to the postal survey and the case studies represent a very skewed sample of local authorities with a bias towards those which, by self-definition, possess integrative activities. However, given that we were attempting to identify the most developed integrative strategies this was not seen as a major problem.

Research results

The relationship between economic-development policy and the environment

In an attempt to discover the extent to which existing policies seek to integrate economic development with environmental policies, postal-survey respondents were asked a series of questions to indicate where integration is taking place. While 71 per cent of Unitary Development (UDP) or Local Plans are said to contain integrative policies, only 47 per cent of Economic Development Plans do so. In assessing how important environmental issues are in terms of economic-development-policy objectives, postal-survey respondents were asked to rank a given number of factors on a scale of 1–10. If these factors are ranked by their mean score, then protecting the environment is not ranked highly in terms of policy objectives (see Table 3).

Despite the relatively limited importance of the environment as a policy objective, a large proportion of local authorities in the postal survey had initiatives in place, or planned, which seek to integrate economic development and the environment. These are shown in Figure 2. Many of these reflect the LGMB's guidelines on greening local economic development, perhaps not surprisingly as most respondents cited LGMB as their main source of information and ideas on such issues. The main initiatives are: sectoral strategies aimed at encouraging the development of environmental-technology firms, both through indigeneous

Table 3 Rank order by mean score of economic-development-policy objectives

Rank order	Policy objective
1	Creating/keeping employment
2	Encouraging inward investment
3	Encouraging new/small firms
4	Costs/availability of land and property
5	Good transport infrastructure
6	Costs/availability of labour
7	Image of the area
8	Protecting the environment
9	Community development
10	Other factors (included housing, city centre regeneration, promoting new technology and sectoral development)

Source: Postal-survey data.

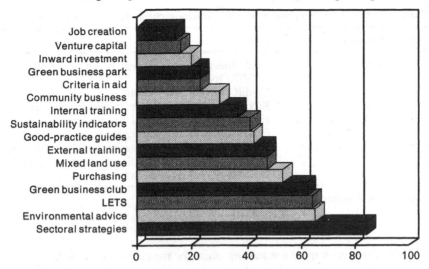

Figure 2 Types of local-authority initiatives linking economic development and the environment (source: Postal-survey data).

development and inward investment; providing environmental advice to firms; establishing Local Exchange Trading Systems (LETS); developing 'green business clubs'; and developing an internal purchasing strategy linked to environmental initiatives.

Nearly half (44 per cent) of postal-survey respondents identified other initiatives in their area which either had no local-authority involvement or are partnership arrangements. These fell into three main categories. The majority of initiatives either provide assistance to, or work with business, or are part of regeneration schemes, usually involving the clean-up of contaminated land or general environmental improvement. A small number of initiatives were community-based, where job or business creation was being linked to the environment (Table 4).

A final section of the postal survey allowed for more qualitative responses on the integration of economic development and environmental policies and initiatives. These began to indicate some of the tensions between the two areas that underlie the simple statements and descriptions of policy initiatives. For example, one comment was that while integrative activities are seen as important, the resources devoted to them often suffer *vis-à-vis* more 'front-line' economic development issues. In part, this was said to reflect the relatively early stage of attempts to integrate policies and initiatives. As one respondent stated 'our Borough Plan and Economic Development Strategy were both drafted some time before sustainability emerged as an issue anyone could understand. They both address sustainability, but not in an overly integrated fashion'. The more narrow view of the environment that has effectively prevailed to date is revealed in a comment that: 'we have environmental improvement strategies – greening the borough – but we haven't got to grips with the sustainability argument'.

Table 4 Initiatives to integrate economic development and the environment

Type of initiative	Examples
Business advice/assistance	Leeds Business Environment Forum ENVIRON, Leicester Cardiff Environment Forum
Regeneration and environmental improvement	Kirkby and Huyton, Town Centre Regeneration Hammersmith, Park Royal Partnership Groundwork Trust, Birmingham
Community development	Miles Platting/Ancoats, environmental job creation Knowsley, Furniture recycling community enterprise Greenwich, recycling electrical appliances

Source: Postal-survey data.

Such comments formed the starting point for the case studies which revealed a diversity of opinion even within those local authorities with integrative policies and initiatives. These interviews suggest that there is a clear split between officers engaged in economic development and those with an environmental remit. In essence there are two interpretations of the environment within local authorities (Myerson and Rydin 1994). For those in economic-development departments, the 'environment' is predominantly taken to mean two main issues. First, it involves the type of relatively small-scale physical improvements to the local environment that can help bolster traditional approaches to economic development, such as inward investment. The 'environment' in these cases is about the 'image of the area', such as landscaping, tree planting and the creation of an 'attractive environment' for potential investors, closely tied in with other regeneration attempts.

A second form of engagement with the 'environment' for economic development departments comes through the need to take account of European policy. In order to obtain European funding, local authorities need to take account of the EU's regulatory requirements (ENDS 1994). For example, Structural Fund Regulations require member states seeking support to produce environmental profiles of the regions in their Regional Development Plans (EU Expert Group on the Urban Environment 1994). The case-study local authorities may comply with these regulations in spirit, but the underlying reality would seem to represent a fairly cynical attempt to secure finance. A disillusioned economic-development officer commented that environmental issues are only mentioned 'when doing so is a requirement for getting money, e.g. European or Department of the Environment grants or budgets, and making "caring for the environment" statements in corporate plans'.

The overall viewpoint that emerges from case-study work with economic development departments is that 'the environment', in any sense, comes a long way down the list of priorities and traditional economic-development activities dominate. A conversation concerning the relative importance of the environment in economic development produced the following interchange between the interviewer and the respondent. This was in a local authority where the economic-development department had produced its own sustainable economic-development policy, which was fully documented with a list of actions and priorities, and which fitted into a corporate 'green strategy' for the local authority.

INTERVIEWER: So how important are the Sustainable Economic Development policies, say compared to Job Creation policies?'
RESPONDENT: They are not ... the overall objective of this department is creating jobs. Full stop. So that anything else we can add to it with sustainability and environmental protection and so on, are a bonus. But the bottom line is creating jobs.

The same respondent spoke at some length on the relative position of the environment in economic development *vis-à-vis* other departments. This gives a clear picture of the 'them' and 'us' mentality that seems to prevail in local authorities.

> Yes, well it's not seen as something that has a high priority, especially within this department. If you speak to other departments you will receive a different answer. Planning for example. They are more sustainability minded. Our remit is to create jobs almost at any cost basically. So if an investor was to set up a factory that was to be polluting but was going to create 500 jobs, our department would argue for that factory to come in. We would have to argue our case to Planning and other politicians and so on. But our politicians from our side would be arguing for it.

Contact by economic-development officers with departments with an environmental remit tended to be extremely limited and few interviewees from an economic-development background would seek the advice of such departments when developing their own strategies. In response to a question on whether the economic-development department would seek guidance in developing their strategy, the respondent replied:

> We might do. If it was a hot issue. But to be honest, that is one of the problems with the Green Strategy ... within this department, is that it doesn't have a very high profile, and it is not seen as being in the top five issues. So if it was higher up the agenda we would do more work on it.

Conversely, departments with an environmental remit frequently adopt a more holistic meaning of the environment. In these cases, 'the environment' is seen as something beyond the local. Environmental policies and initiatives may well contribute to local environmental improvement, but they are also intended to make a contribution to the resolution of global environmental issues. The centrality of the environment for these departments is reflected in the comments of a planning officer upon the UDP process: 'well, from our point of view we have tried to put forward the environment as the lead chapter in the plan'. Environmental officers were much more likely to be aware of (or at least concerned about) the contradictions between economic development and the environment.

The differences of approach are in some cases very clear. An environmental officer commenting on the incorporation of sustainable development issues in their local authority's regeneration strategy document made it clear that whereas the environment group had a clear view of sustainability as involving both local and global contributions, the economic development unit had, in their view, effectively 'hijacked' the word to mean small-scale local environmental improvements, such as landscaping and clean-up of derelict land. Thus, 'sustainability . . . the word featured quite often, but not very effectively, and not in a way that we would understand that term. So it's easy for people to play on words in that sense, to use it as a gloss, rather than having any substance'.

Even in cases where environment officers felt that there was an awareness of the potential to integrate economic-development and the environment within their economic-development department, they were sceptical about the depth of understanding. Again the 'them' and 'us' mentality expressed by economic-development officers also emerges in this context. For example, one comment from an environmental coordinator on colleagues in economic development was: 'I think that they recognize sustainability as a issue, but I don't think they have any understanding or grasp of how they can help the process of working towards greater sustainability. I think that they think that it's tinkering at the edges'.

Officers in environmental departments frequently took the view that economic development within their local authority should pay more attention to alternative strategies, for example: 'my feeling is that rather than traditional economic development, local authorities should be thinking more in terms of the locality of things, and particularly local livelihoods'. From the perspective of environmental officers then there is a perception that the economic-development policies being advocated from within economic-development departments are failing to achieve even their economic aims, let alone contribute to sustainable development. Environmental officers were frequently enthusiastic advocates of the type of initiatives outlined in Figure 2, whereas for economic-development officers these are essentially marginal to the 'main business' of economic development.

Implementation and internal structures

In terms of organization, the postal survey revealed that a majority of local authorities (84 per cent) had an internal committee or sub-committee established to address issues of sustainable development. The financial powers of such committees are limited, with only 25 per cent having a separate budget. These internal committees had representation from four main departments: planning, environment, chief executive's and economic development. However, the dominance of planning and environment departments is shown by the fact that in 64 per cent of local authorities, one or other of these departments is the lead department on the committee (see Healey and Shaw 1994; Owens 1994 for more detail of the role of planning departments). In only 6 per cent of cases was economic development the lead department, although a number of local authorities had joint initiatives.

While such committees are the preserve of officers and members, a number of local authorities had also established bodies to encourage involvement by the wider community. These typically took the form of an Environmental Forum or Local Agenda 21 working party. A major shortcoming of these bodies, from an economic-development perspective, is that most had little or no business representation. In addition to the local authority's own representatives, such bodies were mainly composed of representatives from pressure groups, community groups and residents' associations. This suggests that any private–public partnerships to integrate economic development and the environment currently lack the business involvement that is crucial to their success (Roberts 1995; Welford and Gouldson 1993). Indeed, few case-study respondents involved local business in their strategies. Those that had, discovered apathy and lack of awareness rather than resistance and pointed to the need for a strong economic 'hook' to generate involvement; for example, through stressing potential savings from greater energy efficiency or energy reduction. Greater success in working with business from an environmental perspective was found by those working in economic-development units. Business was said by these units to view officers from environmental departments as a potential threat, with fears of reporting back to the council on environmental infringements.

Attempts at integrative activity varied considerably. The case studies revealed that some councils have paid particular attention to developing projects and strategic policy in parallel. In Leeds and Southampton integrative policies appear in corporate-wide environmental strategy plans which are intended to drive strategy and initiatives in individual departments. For example, Leeds City Council have a comprehensive set of documented policies to integrate economic development and the environment in the council's green strategy action plans. Each department of the council is required to generate and implement action plans detailing the actions to be taken at departmental level in support of the council's green strategy. The Leeds Green Strategy Review outlines the original objectives and then reviews them in terms of 'position to date' and 'objective status'. While on paper these policies appear to be well developed, the value of such strategy plans, in terms of the weight they carry, can be questioned. A closer look at the Leeds UDP and economic-development strategy reveals that although the green strategy is referred to, the advanced level of policy set out in the action plans is not duplicated in these plans. Conversely, Bradford, where environmental issues form part of the economic development strategy, is more advanced in terms of having set objectives and programmes that can be defined as integrating economic development and the environment. It could be argued that such policies may be more effective than policies established as part of a corporate environmental strategy in that they have been set by the economic-development department itself, rather than imposed upon them. In Cardiff, Manchester and Kirklees environmental priorities are largely set by either the planning department or the environment unit respectively, with relatively little impact upon the economic-development departments.

Just under half the postal-survey respondents (44 per cent) undertake any environmental evaluation of their economic-development policies. Even here, however, the means of doing so tended to lack any formal mechanism, with 48 per cent using the existing planning-appraisal process by reference to the local plan, and a further 20 per cent who said they used 'personal judgement' without specifying the process. Few used the Department of the Environment's 'good practice' appraisal guide or any form of strategic/environmental-impact assessment (Department of the Environment 1993). Even in the case studies, few engaged in monitoring the impact of any initiatives undertaken, particularly in terms of the environmental impact of economic development. Even where this did occur the evaluation was cursory, unsystematic and unquantified. Respondents in Kirklees and Manchester were in the process of developing methods of 'sustainability assessment' while Cardiff were exploring the potential to use indicators of sustainability. Any evaluation that does occur did not appear to be using any set methodology. Again, the most common example involves authorities relying on their own judgement or using the guidance provided by the national environment ministry.

One issue which emerged from the case studies is the importance of high-level support for strategies within the local authority. A committed leader of the council appears to make a substantial difference to implementation and engenders internal support and access to finance. In the absence of high-level support environmental initiatives are frequently seen as marginal to the main business of the local authority, particularly in terms of economic development. This may mean that environmental issues are confined to environmental departments or units, rather than being part of corporate policy. One respondent commented: 'you have to get ownership of the environment as a corporate issue at the highest levels in the council. And so getting senior officers and executives on board, aware of what the implications are and to accept that it is something that permeates all levels and isn't just a project'. The case studies indicated numerous examples of the problems of cross-departmental working whatever the internal local-authority structure. Even in instances where a corporate environmental initiative had been established to avoid problems of compartmentalization, such as in Kirklees, there was still evidence of environmental issues being seen as the sole province of the Environment Unit. This suggests that strategy co-ordination is a key area for improvement. The case studies indicated greater success in integration where environmental issues had been reinterpreted into an appropriate context. As one respondent commented: 'once people latch on that LA21 and sustainability were about quality of life, they started to realize how their strategies and policies fitted into it'.

Postal-survey respondents were asked what they saw as the major obstacles within their local authority to implementing or furthering initiatives which try to integrate economic development and the environment. Of those factors listed as most important, lack of finance was the dominant reason, followed by the fact that other priorities, such as jobs and firm closures, were deemed to be of more importance. Obviously not all problems with regard to integrating economic development and the environment can be assigned to internal problems. Two themes which emerged from the case studies were a lack of guidance on what constitutes a sustainable local economy and the lack of commitment shown by central government, despite the rhetoric on sustainable development. Concrete evidence of the latter was shown in the problems several case-study authorities had with their unitary development plans. Including sustainable development aims in the UDP had been criticized by the Department of the Environment as not being a statutory requirement. Such DoE advice was criticized by case-study respondents as 'confused and contradictory'.

Conclusions

One of the major aims of the research was to investigate the extent to which local authorities are integrating economic-development strategies with their environmental strategies. A substantial body of evidence already existed which outlined the rise of environmental strategies and policies within local authorities (Agyeman and Evans 1994; Tuxworth and Carpenter 1995). However, our initial reaction to these was that most were based upon either a fairly simple reading of publicly available documents or upon survey work which rarely investigated the subject in depth, merely noting the presence or absence of a strategy or documentation (Gibbs 1993). Our aim was to go behind simple statements of intent which indicate a high level of local-authority activity on integrating the environment and economic development and to examine the implementation of related policies and strategies and the political importance attached to these internally. As this chapter reveals, taking local-authority statements on levels of activity at face value can be misleading. While there has been a move towards integration of these two policy areas, as yet this has occurred at a fairly superficial level and is not widespread across UK urban local authorities. Moreover, one of the important findings from the case studies was that even in local authorities where integration appears to be well developed there remain different interpretations of what constitutes 'the environment' between practitioners in economic development and environmental or planning departments. In part these problems arise through the effective continued separation of the two functions within local authorities. Even in cases where attempts had been made to integrate the two areas, economic development appeared to dominate, such that the integrative policies and initiatives form a sub-set or 'add-on' to the main business of economic development. Under such a regime, the environment in a local economic-development context is becoming just another component to be packaged for corporate consumption and inward investment purposes (Gibbs 1997). There is still a need for improved strategy coordination within local authorities and for placing environmental issues at the heart of local economic-development policy.

In terms of insights into the possible development of a framework to implement sustainable city strategies, barriers to achieving a workable framework for implementing sustainable city strategies were found to be both theoretical and practical. In the case of the former, case-study local authorities did not have a clear vision of how a sustainable local economy could be achieved. Available guidance was seen as existing at a high level of abstraction resulting in vague recommendations, lacking in practical application. One consequence of this is that even in local authorities with a keen interest in sustainable development, the initiatives taken are piecemeal rather than forming part of a holistic, authority-wide response. There is a need to develop more 'visions' of what a sustainable city could, or should, look like and exchange of good practice between local authorities (see Ravetz 1996 for an attempt at the former). In the case of practical barriers the perceived lack of commitment from central government was a hindrance, compounded by the limited funding available to implement initiatives. Internally, in the case of economic-development officers, other priorities such as jobs and firm closures were deemed of greater importance. Practical barriers also exist in the internal structure of local government. The case studies threw up differing models of implementation, but the dominant trend is for economic development and environmental functions to remain separate areas of responsibility within local authorities. Integrative strategies were much less advanced than we had envisaged at the outset of the research and examples of existing 'best practice' were subsequently lacking.

In total then we obtained very limited evidence of integrative activity and, more importantly, two very different views on the role of the environment within local authorities. This leads us to the conclusion that until these views are reconciled there is little prospect of truly integrationist approaches. To integrate economic development and the environment would require methodologies for appraising and assessing proposals and plans and prioritizing developments that do not, as yet, exist. At present these are difficult to conceive of in a situation where local issues are paramount and short-term politics dominate. A longer-term integrationist approach can be envisaged when new entrants to the local-authority professions have undergone education in these issues and start from a premise of holistic assessment. There is a role for continuing professional development within both economic development and environment departments to bring together these perspectives and to resolve the tensions. This suggests that although top-down policy from national and EU policy-makers assumes a fairly unproblematic integration of economic development and the environment occurring at the local level, the research suggests that this is far from the case (Stren 1992; Keil 1995). Local authorities may well need to alter their own internal structures and thinking, but they also need to be given adequate funding, powers, training and an appropriate policy context for their actions. Without these changes the future prospects for integrating economic development and the environment will remain fairly bleak. For local authorities such changes would provide potential to develop a set of new and refocused economic-development policies which place the environment at the heart of economic development.

These could include: targeting inward-investment policies on environmental-technology sectors; encouraging improved environmental standards through supply chains; demonstration projects on waste management and pollution control to local firms; developing local exchange-trading systems (LETS); creating jobs through environmental-improvement schemes; and devising local indicators of sustainability for such policies. While some of these developments are in evidence in a piecemeal fashion, in future they need to be part of an integrated holistic strategy.

Acknowledgement

The research on which this paper is based was funded through the ESRC's Global Environmental Change Programme (grant number L320252132).

Endnotes

1 These bodies were identified from the Municipal Yearbook as the major players in local-government policy and practice. They were: Association of Metropolitan Authorities; Association of County Councils; Association of District Councils; Association of Greater Manchester Authorities; Association of London Authorities; International Council for Local Environmental Initiatives; International Union of Local Authorities and Councils: British Section; Local Government International Bureau; Local Government Management Board; London Boroughs' Association; London Planning Advisory Committee; London Research Centre; National Association of Local Councils; and the Northern England Assembly.
2 The postal survey was sent to all sixty-nine English metropolitan authorities (London boroughs and metropolitan boroughs), plus all local authorities in England and Wales designated as 'urban' on OPCS criteria.
3 Interviews were undertaken within City of Cardiff Council, Bradford Metropolitan Council, Kirklees Metropolitan Council, Leeds City Council, Manchester City Council and Southampton City Council. Secondary information, such as Unitary Development Plans, Economic Development Strategies and Environmental Strategies and other relevant policy documents were obtained from these councils, together with documentation from other bodies external to the local authorities.

References

Agyeman, J. and Evans, B. (1994) *Local Environmental Policies and Strategies*, London: Longman.
Agyeman, J. and Evans, B. (1997) 'Is anyone listening?', *Local Environment* 2(1), 5–6.
CEC (Commission of the European Communities) (1992) 'Towards Sustainability – a European Union programme of policy and action in relation to the environment and sustainable development, COM 92(23), Brussels: Commission of the European Communities.
DoE (Department of the Environment) (1993) *The Environmental Appraisal of Development Plans: A Good Practice Guide*, London: HMSO.

ENDS (1994) ' "Screening" and "scoping" proposed for environmental assessment', *ENDS Report* 232, 38.

EU Expert Group on the Urban Environment (1994) 'European sustainable cities', first report, Sustainable Cities Project, XI/822/94-EN.

Gibbs, D. C. (1993) *The Green Local Economy*, Manchester: Centre for Local Economic Strategies.

Gibbs, D. C. (1997) 'Urban sustainability and economic development in the United Kingdom: exploring the contradictions', *Cities* 14(4), 203–8.

Gibbs, D. C. and Healey, M. J. (1995) 'Local government, environmental policy and economic development', in M. Taylor (ed.) *Environmental Change: Industry, Power and Policy*, Aldershot: Avebury, pp. 151–67.

Gibbs, D. C., Longhurst, J. W. S. and Braithwaite, C. (1996) 'Moving towards sustainable development? integrating economic development and the environment in local authorities, *Journal of Environmental Planning and Management*, 39(3), 317–332.

Hams, T. and Morphet, J. (1994) 'Agenda 21 and towards sustainability: The EU approach to Rio', *European Information Service*, 147, 3–7.

HM Government (1990) *This Common Inheritance*, Cm1200, London: HMSO.

HM Government (1994) *Sustainable Development*, Cm2426, London: HMSO.

Healey, P. and Shaw, T. (1994) 'Changing meanings of "environment" in the British planning system', *Transactions of the Institute of British Geographers*, 19, 425–38.

Keil, R. (1995) 'The environmental problematic in world cities', in P. Knox and P. Taylor (eds) *World Cities in a World System*, Cambridge: Cambridge University Press, pp. 280–97.

LGMB (1993) *Greening Economic Development*, Luton: Local Government Management Board.

Myerson, G. and Rydin, Y. (1994) ' "Environment" and "planning": a tale of the mundane and the sublime', *Environment and Planning D: Society and Space*, 12, 437–52.

Owens, S. (1994) 'Land, limits and sustainability: a conceptual framework and some dilemmas for the planning system', *Transactions of the Institute of British Geographers*, 19, 439–56.

Roberts, P. (1995) *Environmentally Sustainable Business: A Local and Regional Perspective*, London: Paul Chapman Publishing.

Ravetz, J. (1996) 'Towards a sustainable city region', *Town and Country Planning*, 65(5), 152–4.

Stren, R. (1992) 'Conclusion', in R. Stren, R. White and J. Whitney (eds) *Sustainable Cities: Urbanization and the Environment in International Perspective*, Boulder, CO: Westview Press, pp. 307–15.

Tuxworth, B. and Carpenter, C. (1995) *Local Agenda 21 Survey 1994/5: Results*, Luton: Local Government Management Board.

UNCED (1992) *Agenda 21*, Conches, Switzerland: United Nations on the Environment and Development.

Welford, R. and Gouldson, A. (1993) *Environmental Management and Business Strategy*, London: Pitman Publishing.

7 The interpretation and implementation of sustainability strategies in UK local government

Stephen Rees and Walter Wehrmeyer

Introduction

In the UK, the integration of environmental management and regional and local economic development has gained much interest following the advent of sustainable development (SD). The landmark in this process has been the introduction of Local Agenda 21 (LA21), now the most significant local-authority contribution to the commitments made at the Rio Earth Summit in 1992. In many respects, LA21 is an attempt to stimulate community governance and local democracy so as to encourage accountable decision-making and the fair and appropriate use of local and global resources. However, measures to promote forms of development that meet the needs of the current generation without sacrificing the ability of future generations to meet their needs – the famous Brundtland definition of sustainable development – have largely been adopted by UK local government in the absence of additional resources or a discernible strategic approach. This has had obvious implications for the capacity of local authorities to pursue integrated approaches to environmental management and economic development under the 'sustainable development' umbrella. It has also affected the nature and types of response that local authorities have taken.

Initially, this chapter reviews the integration process by means of a survey of the Local Government Environmental Co-ordinators (LGECs) that are responsible for the promotion of SD both within the internal structures and operations of particular local authorities and through the policies and plans that these authorities develop and implement. The semi-structured questionnaire upon which the first part of this chapter is based was sent to 453 LGECs in 1995. The 153 responses were both geographically and organizationally representative and the main methods of analyses were consistent with existing appraisals to attitude surveys using basic cross-tabulations and Principal Component Analysis. Interestingly, of the 153 respondents, 17 per cent were planners by job title, 37 per cent were environmental health officers, 11 per cent were in policy-making and only 24 per cent were environmental co-ordinators. This suggests that a wide variety of backgrounds and perspectives are involved in the operationalization of SD at the local level.

The results suggested that when the survey was conducted in 1995 local authorities in the UK were generally ill-equipped to meet the challenge ahead due to poorly designed organizational and regulatory frameworks. Irrespective of the structural limitations, the survey also found that LGECs encountered significant limits to financial and human resources within their own authorities that prevented them from developing new initiatives and exerting their influence fully. Thus, at the time of the survey the response of local authorities in the UK to the challenge of sustainable development was in grave danger of being either malformed or strangled at birth. Local authorities in the UK, locked into their statutory roles and responsibilities, were having severe difficulty finding the resources to sustain the *status quo*, let alone finding the resource base with which to create sustainable communities supported by a sustainable local economy and environment. While the situation has obviously evolved to some degree since the survey, notably in relation to the accumulating experience with implementation and the emergence of new networks to encourage and transfer good practice, there is little evidence to suggest that the current situation is radically different from that revealed by the survey in 1995.

Following this discussion, the second part of this chapter advocates a range of changes to the values, objectives, mechanisms and processes adopted by local authorities as they seek to promote meaningful change that is compatible with the broad concepts and philosophies of sustainable development. An illustrative 3-year strategic plan for sustainability is also proposed. The chapter concludes by suggesting that although significant progress has been made in some respects in recent years, as yet there is little evidence to suggest that the sustainable development agenda has had a radical impact on the strategies and structures of many if not all local authorities in the UK.

Recognizing the need for change

LGEC interpretations of SD

Given the ambiguities associated with the definition and application of SD, the research specifically asked LGECs for a definition of this contentious and ambiguous concept. This was done not only to identify the operational dimensions of the term but also to see what particular emphasis and preference was taken in the integration of environmental management and economic development. In the survey, 41 per cent of LGECs defined SD using the convenient Brundtland definition (WCED 1987). However, the majority of LGECs still preferred to define SD in their own terms (the italicized text below refers to quotes from questionnaire survey results), including the definition of one environmental co-ordinator who stated that:

> *Sustainable development means carrying on with anything you like but justifying it with clever arguments.*

More generally, LGECs described SD in the past context by continually referring to *irreversible environmental damage and degradation, wants greater than needs, short-term decision-making,* and *the inequitable and destructive misuse of resources.* In addition LGECs defined SD in the present in terms of identifying the *needs and wants of today, natural and human capital/stock, ecosystem and cultural system carrying capacities.* These descriptions revealed a deeply felt concern by respondents over the current state of the environment and human interactions with it. The descriptions depicted a global and local society locked into unsustainable forms of economic activity. These feelings were largely personal ones which somewhat differed from the feelings and values espoused by LGECs in their professional lives. This supported the research of Glasser *et al.* (1995) on the values held by senior policy advisors in the European Union. He found that even though most 'articulated deeply held personal environmental values, (they) normally keep these values separate from their professional activities' (p. 83).

Somewhat differently from their interpretations of SD in the past and present, the future was couched in much more positive terms – *a diverse and healthy environment, a future for our children, an improving quality of life, sustainable communities, a new world economic order, local and global equity, common sense, well being* and *cultural tolerance.* What was missing from this veil of optimism was a consideration of the mechanics and the process with which SD might be operationalized or implemented.

Many LGECs confirmed the problem in their own organizations and in their own communities of individual preference being at the expense of collective responsibility. The current economy was seen as supporting and encouraging an unsustainable approach, driven by unrestrained human behaviour, lifestyles and consumption patterns. In fact, this problem of allocating collective resources according to individual criteria (and the resulting dysfunctional effects) has been discussed for some time, notably since Hardin's classic paper on the tragedy of the commons in 1968 (Hardin 1968). It also played an important role here. For instance, 88 per cent of LGECs were against the current global approach to free enterprise and market deregulation. There was a perceived need for the 'free market' of options and freedoms (supply and demand) to be better defined within local and global social and environmental limits and carrying capacities. However, 71 per cent of LGECs thought that local authorities had a lack of influence on patterns of economic production, distribution (trade/markets) and consumption, making the aspired integration of economic development and environmental management substantially less likely. However, the chances of fundamental changes in consumption patterns were seen as slim.

Ironically, when LGECs were asked to rank their SD goals in order of priority, *the regulation of a supply-led market-driven economy* (rank 9) was rated well below *preventing human poverty and suffering* (1), *maximizing education and awareness* (2), *moving towards renewable production and consumption patterns* (3), *setting up social and cultural systems which reward sustainable behaviour* (4), *minimizing the loss of biodiversity* (5), *addressing climate change* (6), *addressing ozone depletion* (7), *moving towards zero waste and pollution* (8).

In relation to the identification or introduction of environmental limits, at the time of the survey local authorities in the UK were only in the early stages of trying to define those 'limits' by updating their planning, development control and contractual mechanisms and by attempting to construct better measures of economic, social and environmental well-being through identifying primarily local sustainability indicators. The results indicated that 51 per cent of LAs had undertaken environmental appraisals of development plans, 46 per cent were developing environmental reporting mechanisms (environmental management and sustainability indicators of performance), 38 per cent were integrating environmental specifications into contracts, 37 per cent had conducted state-of-the-environment audits and 66 per cent were developing environmental education programmes.

The role of local authorities

Table 5 illustrates five core areas of activity derived from LGEC survey responses. The broad concepts of care, collaborate, create, communicate and control represent the key functions of many if not all local authorities. These broad concepts of what a local authority does, have in turn been under relentless stress in recent times.

Local authority activity in the UK has been dominated by great organizational change ranging from the introduction of compulsory competitive tendering and local government reviews, changes in strategic funding, severe public-expenditure cuts and increasing levels of new legislation. In 1998 the threats and uncertainty of the last decade continue for local authorities with the creation of regional development agencies and moves towards new regional government. In the short term this will not give local authorities a stable basis from which to pursue SD. In the longer term the devolution of power from central-government should lead to greater control and direction at a regional and local level. The survey found that 82 per cent of LGECs considered central-government financial and statutory arrangements to be a problem in achieving sustainable development. This broader context limited the time and resources available to local authorities to address the sustainability agenda let alone take any significant steps to encourage the

Table 5 Local-authority role as perceived by LGECs

CARE	As defender, provider, carer, steward and protector
COLLABORATE	As enabler, facilitator, participator and partner for projects, ideas and activities
CREATE	As initiator, example setter and educator for pilot and ongoing activities based on best practice
COMMUNICATE	As lobbyist, awareness raiser, arbitrator, conflict resolver and informer
CONTROL	As appraiser, auditor, planner, designer, manager, monitor and regulator within defined economic, social and environmental limits

development of sustainable local communities and economy. This was com-pounded by the fact that many of the national initiatives demanding change (and therefore creating organizational instability) in local government had not genuinely attempted to adopt sustainability criteria as part of their remit. This perhaps reflects the limited degree to which some central-government depart-ments and quangos had truly grasped the concept of sustainability and under-stood how best it could be delivered at a local level through local-government machinery.

Concepts of local leadership or governance (to educate, direct and regulate) may sometimes appear to be a forgotten art form but they are slowly re-emerging via a recognition and appreciation of the importance of the environment alongside issues relating to social and economic well-being. However, many LGECs expressed a feeling of unease that local authorities might ignore, neglect or feel unable to address the opportunities available in each of these areas through the SD agenda. Alongside fears related to the organizational distractions imposed by central government were also more immediate worries about the narrow interpretation of SD that was commonly adopted. In many local authorities it appeared that SD had continuously been referred to in a physical context within planning, highways, economic development and environmental health functions. The development of social or cultural values and behaviour to influence resource use and consumption patterns had commonly been overlooked or had become secondary.

The survey also highlighted a distinct gap in skills and resourcing on the social and educational side, an area which many see as being of crucial importance for the success of LA21 and community development programmes. The predomi-nance of the physical planning and management professions in constructing the local-authority agenda does not necessarily bode well. These professions, traditionally dominated by specialists in environmental health, land-use planning, civil engineering and development, were consciously or unconsciously excluding the social practitioners such as community developers, social scientists and educationalists who arguably are better capable of identifying social needs and wants and helping to design better physical infrastructure and socio-economic frameworks to support sustainable communities and lifestyles.

The survey found that 62 per cent of LGECs considered the 'capture' of the SD agenda by particular professional groups as being a problem as it disempowered communities from constructing and owning social, economic and environmental change. The 'professionalization' of what needs to be a democratic and participatory process was of concern to many LGECs, some of whom even question the nature of their own emerging environmental profession, fearing that it would create an exclusive rather than an inclusive professional discourse as well as professional structures dedicated to defending the economic position of the profession itself rather than promoting the interests of those it was established to serve.

The survey also suggested that this perception of a 'professional' takeover of the SD agenda was commonly exacerbated by the current means of communication and consultation through plans, reports and literature which are highly technical and legal which often exclude the disadvantaged, less articulate or illiterate. In fact, 71 per cent of LGECs believed the emerging professional language and terminology used to express environmental impacts and actions was a problem. The importance of clear and accessible language and exciting media to promote simple principles, goals and indicators was highlighted by many LGECs. Similarly, the survey indicated that the professionalization of the area prevented LGECs from enacting their environmental beliefs for fear of being seen as naive, unrealistic and not fitting into the dominant mode of economic thought.

A further dilemma facing local government, as it sought more constructive approaches to SD, related to the issues of legitimacy and credibility. The survey established that many LGECs perceived themselves and government generally to be mistrusted by the community, by Non-Governmental Organizations (NGOs) and by businesses. It suggested that their current image was compounded by a historically poor perception of local politicians and of the systems of government that promote politically expedient objectives in the short term rather than social, economic and environmental objectives in the medium to long term.

Barriers to the adoption of SD as a policy objective

As suggested in Table 6 below, following an analysis of the survey results, it appeared that there were many barriers preventing the adoption and implementation of SD as a policy objective by local authorities in the UK.

First, it appeared that there was an issue about the definition and understanding of SD as it remained a vague concept for those charged with adopting and implementing policies to promote SD. Second, it was evident that those organizations and departments charged with facilitating a shift toward SD perceived this task to be of less importance or relevance than their traditional

Table 6 Barriers to the adoption and implementation of policies to promote SD by local authorities

---

- 76% of LGECs recognized a lack of public awareness and understanding about what sustainability was about
- 81% of LGECs recognized that SD was not the top priority in their authorities
- 71% of LGECs stated that there was a lack of understanding of the concept of SD by managers/politicians
- 71% of LGECs pointed to a lack of an appropriate structure in their local authority to adequately confront the real issues
- 75% of LGECs also stated that they lacked the necessary time and resources to confront issues of organization or action adequately
- 72% of LGECs stated that there was a lack of environmental management capacity in their organization
- 82% of LGECs expressed great concern about the short-term nature of political systems and 81% of internal financial arrangements
- 65% that there was inertia and resistance to change

---

roles. Third, it was apparent that concerns existed about the limited ability of local authorities to influence the generation, regulation and redistribution of local resources and to introduce novel and radical concepts and processes into the local community and economy. This perceived lack of influence was compounded by the lack of stability in local government and by significant resource shortages which limited the potential for creative investment to catalyse or support good practice in the voluntary, private and public sectors. It was also compounded by the apparently limited commitment to SD in central government, by the perceived poor reputation of local government and by the feeling that local authorities operated within a climate of fear and insecurity which was not seen to be conducive with innovation and leadership. Fourth, the short-termist nature of political decision-making was highlighted as a significant barrier to the adoption of policies that could contribute to a shift toward SD. This was seen to be the case as political representatives and personnel were perceived to be locked into short-term political and financial cycles which, along with various legal constraints, severely restricted their ability to take radical decisions and a more long-term view.

Further discussion of survey findings

The research findings therefore suggested that SD initiatives have been adopted against a background of apparently limited top-level commitment, scarce resources, inappropriate organizational structures and restricted influence. More positively, and partly as a response to these less-than-ideal circumstances, it was encouraging to see the development of a cohesive network of LGECs sharing best practice and experience. The emergence in 1997 of a new central government apparently committed to regional devolution and the promotion of sustainability was also a promising development which had the potential to help to solve the various problems that some actors within local authorities face as they attempt to promote SD. In their attempts to operationalize LA21, there is a good chance that more robust organizational values, long-term service strategies and medium-term business plans will evolve in local government to underpin the wider adoption and evolution of sustainable development and LA21 programmes. However, arguably there are three more areas of concern.

First, LA21 was and still is an opportunity for LAs to reassess their role in the communities they serve. This opportunity appears to have been largely lost in many authorities that have incorporated SD into their existing organizational frameworks without involving the public, voluntary agencies and business in a sincere and challenging review of the local environment and of the socio-economic context for promoting sustainable development in practice.

Second, since organizational capabilities have been severely reduced within LAs over time, the capacities of LAs to deal with SD is substantially hampered by the wide range of current statutory demands. The willingness to take on a whole new agenda that is essentially voluntary is therefore seen as an extra demand on already over-stretched resources rather than as a challenging opportunity. The priority assigned by central government to the promotion of SD at the local level is therefore of great importance.

Third, organizational instability, the lack of resources and job insecurity, coupled with the crisis of identity in local government, has fostered greater reliance on those resources allocated to a department for particular programmes and initiatives. This has resulted in interdepartmental pressures to claim owner ship of the SD/LA21 process as an issue to be addressed by planners, environmental specialists or other professional groups as a way of defending themselves against organizational change and budgetary cuts. However, sustainability is essentially a lateral and integrative concept that addresses all levels of an organization, economy, environment and society. Integrated approaches to planning, policy-making and implementation and resource management are therefore vital both within organizations and across the local economy. The emergence of an 'us' and 'them' approach to sustainability, whether within a local authority, between a local authority and its various constituents or between various social groups is problematic. This is perhaps a symptom of the vagueness of the concept itself and the increasingly indistinct and uncertain role of a local authority and its various functions.

The discussion above indicates that there is a need to adopt an integrative and participatory approach based on strategic vision and social partnerships. However, it has also indicated that the ability of LAs to adopt and implement policies that are compatible with SD may be limited. Without appropriate powers, sufficient financial resources, committed and educated personnel and a significant reorganization of the machinery of local government, it is unlikely that LAs can make a contribution worthy of their role. There is a great and fundamental danger that through LA21 and other SD initiatives local authorities will raise these aspirations, which can somehow reinforce the low opinion the public often has of their local authority.

Strategies for change

Given the discussion above, it is appropriate to consider how local authorities with limited resources and relatively underdeveloped capacities for change can establish policies and programmes to deal with the SD/LA21 agenda. In this respect, it is apparent that a local authority must have the democratic 'authority' to lead and govern within its own region if it is to promote progress towards sustainability. In this respect, a number of core values must first be identified and translated into organizational change and action to create and validate an appropriate approach to government at the local level. Examples of such core values are set out in Table 7.

Upon the basis provided by these values, action can be taken to further develop an authority's capacity to promote long-term and continuous improvements in environmental, social and economic well-being in the region. A number of common challenges are apparent in this respect.

Table 7 A strategy for sustainability – the core values of local authorities

---

Eco-efficiency and quality services
Making the best use of limited resources to deliver effective, accessible services to meet essential needs

Local democracy
Developing and improving the democratic role of local government in supporting community governance and participation in the engineering of sustainable communities and a sustainable local economy, defined within social and environmental limits (carrying capacities) – local and global

Priority for those most in need
Giving priority to policies and services which redress disadvantage and promote opportunity and fairness

Safeguarding the future
Adopting a long term and sustainable approach which enhances quality of life for human and natural communities both locally and globally

---

The challenge of integrated resource management

Many of the authorities that have been subjected to extreme financial pressures have in recent times become more resource efficient. Commonly, resources are being used more productively. However, this does not mean that resources are being well targeted. In this respect it is evident that financial management within local authorities has traditionally been dedicated to cost–benefit analysis in its narrowest economic sense. This has limited the ability of many authorities to take less tangible social and environmental costs and benefits into account. In addition, as public aspirations are raised on social and environmental issues, decisions made on purely financial grounds are becoming increasingly difficult to justify. New government proposals for 'best value' may present a more viable opportunity for LAs to reflect sustainability objectives (social and environmental costs and benefits) in their decision-making – particularly on financial and purchasing decisions.

The question therefore arises of how authorities can alter their own financial management practices in order to channel their resources more effectively so that they contribute to the core values set out above. In many instances this is likely to require changes in financial management and accounting practice so that it shifts from its 'neutral' stance to find creative ways to accentuate positive action (towards sustainability) and minimize negative action (away from sustainability) both within and outside the authority.

Dealing with the symptoms of unsustainability

Many of the functions of local authorities are driven by statutes and traditional activities which are often reactive in nature. In many instances, authorities are obliged to deal with the after-effects of unsustainable social and economic activity.

The shift towards a more preventative strategic approach will require a transition in resource allocation and a move away from reactive and costly public services which deal with problems rather than solutions. The moves towards a more preventative approach to tackle issues at source are already beginning to take place in local authorities: For example, crime and fire prevention; community development; sustainable design in planning; development and regeneration projects; preventative health care; environmental, LA21 and sustainable business programmes; eco-efficiency programmes; lifelong learning in education; waste minimization.

However, many of these activities are limited, under resourced, sporadic, fragmented and unable to fundamentally transform damaging social and economic activity on a large scale. It is rare that LAs are given the powers, resources and competence to change the very nature of their activities to eliminate or at least minimize undesirable impacts while optimizing good practice. The intensity of activity required to shift a community or economy from unsustainability to sustainable is sadly missing from government activity at a local, regional and national level.

Local authorities can become more proactive by promoting greater participation and by fostering partnership with the various actors in the local economy and community so that they can influence the nature of development rather than merely responding to its negative impacts. Through such partnerships, local authorities can extend their influence to catalyse or encourage more sustainable patterns of production, distribution and consumption whilst at the same time relieving pressure on traditional local-authority-resource commitments. For example, local authorities can develop community action, self-help and structures to support disadvantaged social groups, they can promote local clean production, they can encourage local self-sufficiency to minimize need for transport or waste management, they can establish opportunities for participation in the planning process to empower local communities and improve the outputs of the planning process, they can promote environmental awareness and education to change social values and behaviour patterns, they can support local environmental initiatives to improve the local quality of life and so on.

The power, influence and resources of local authorities

It is vital that local authorities do not get 'locked-in' to financial and legislative mechanisms that only 'sustain the unsustainable'. However, it is clear that the ability of individual local authorities to influence the over-arching structures that guide the development process is limited. Consequently, moves to align international, national, regional and local government strategies and priorities are needed to ensure consistency and continuity and to avoid duplication and conflict. In this respect there is commonly a need for a clearer division of responsibilities as the ability of local authorities to promote an agenda that challenges that supported or accepted by regional, national and supra-national tiers of government is limited. In many instances in the UK, a more explicit definition of the role and remit of local authorities, particularly with regard to their regulatory, fiscal and economic powers, is also important as uncertainty inhibits the promotion of clear strategies to promote the SD agenda.

The way in which resources, including people, information, material and finance, are targeted is also of great significance. As mentioned above, local authorities can extend their influence where they build partnerships with the various actors and groups that shape the nature of development within their localities. However, local authorities can also extend their influence by targeting their activities, and those of the partnerships within which they operate, so that they influence local people at key points of cultural transition in their lifetime. From this research, we know that these key points include events such as when young people start and leave school, when they start or change jobs, move house, start families and so on. Cultural studies suggest that peoples' preferences and behaviour are more dramatically influenced at particular moments in their lives than at others. Therefore seeking to raise the level of awareness of more sustainable opportunities at these moments may have a more significant impact over the longer term. In particular, media campaigns, education and incentives and disincentives can promote sustainable lifestyles especially if they are targeted at the key moments of transition in the personal development of the individuals who collectively constitute the communities that local authorities attempt to serve.

As with external finances, the internal finances of many local authorities can be locked into set patterns of investment which are not necessarily sustainable. Commonly, there is competition between the objectives and management systems of the various departments, between professional and resource-management functions and between service providers (business units) that is not conducive to the implementation of an integrated strategy. Consequently, resources are not generally allocated or prioritized against sustainable criteria. This reflects the fact that many authorities are organizationally fragmented, task/ procedure driven and short-termist in their outlook and approach. Thus, choice in how resources are used effectively and creatively to obtain long-term 'value for money' is constrained.

To rectify this situation, many authorities need to restructure their organization and lengthen their business planning and financial cycles, particularly by establishing devolved targets, budgets and performance measures, to enable managers to reassess their perception of 'power', 'culture', 'objectives', 'choice', 'success', 'efficiency' and 'value' over the long term. An understanding of social and environmental costs and benefits also needs to be embedded to enable decision-making to be made more effective. In all respects, many authorities need to reassess their values, functions and activities so that their strategies, which should reflect the central concepts and goals of sustainability, are well designed and their operations actually meet their longer-term objectives.

The strategic criteria that provide the basis for decision-making in many local authorities need to become more transparent over time so that authorities can demonstrate moves towards more sustainable forms of resource management. For this to happen, elected members will need to recognize that the decision-making process as well as its outcomes will need to be made more transparent and accountable. Opportunities for public participation in decision-making processes can also be usefully established so that decisions contribute to the realization of public aspirations as well as the long-term strategic objectives of the authority itself.

Conclusion

This chapter has suggested that local authorities in the UK have yet to demonstrate that they have the power, the capacity or in some instances the credibility to move beyond their current approach which tends to manage some (but not all) of the symptoms of an unsustainable society and economy. Given the extent to which power has been centralized in the UK in recent decades, and the severe pressure that has been put on the resources allocated to local authorities to fulfil their statutory responsibilities, the capacity of many local authorities to promote sustainable development effectively is in severe doubt. The results of the survey presented in this chapter suggest that local-authority initiatives to promote more sustainable forms of economic activity tend to be fragmented in their design, limited in their scope and marginal in their impact.

Through initiatives such as LA21, some environmentalists have at last begun to move beyond the historical doom-and-gloom predictions to confront the real issues at source and to attempt to overcome the barriers precluding progress. It is perhaps time that the politicians and public servants who formulate the strategies of local government attempted to re-constitute the way in which they look at the world rather than sustaining the *status quo*. In this respect, it is apparent that the SD agenda needs to alter patterns of production and consumption and to influence public awareness and behaviour rather than merely protecting and accommodating unsustainable forms of social and economic activity.

However, the extent to which existing local-authority structures and activities depart from the sustainable ideal is significant. This chapter has essentially argued that incremental steps towards sustainable development within current structures and systems are not possible or advisable. New structures and systems are needed, some of which must be introduced at the international and national levels. At the local level clearly there is a vital role for initiatives that bring together the various actors and interest groups to establish the values, objectives, mechanisms and processes that are needed if the broad concepts of the sustainability agenda are to be implemented practically. Although significant progress has been made in some respects in recent years, as yet there is little evidence to suggest that the sustainable development agenda has had a radical impact on the strategies and structures of most if not all local authorities in the UK.

Acknowledgements

This research enjoyed the support and endorsement of the UNED-UK

Committee, the International Council for Local Environmental Initiatives (ICLEI), the Local Government Management Board (LGMB) and the Environment Council, UK. The authors wish to express their gratitude to these organizations for their contributions.

References

Glasser, H., Craig, P. and Kempton, W. (1995) 'Ethics and values in environmental policy: The Said and the UNCED', in J. van den Bergh and J. van der Straaten (eds) *Toward Sustainable Development*, Washington: Island Press, pp. 83–106.

Hardin, G. (1968) 'The tragedy of the commons', *Science* 162, 1243–8.

WCED (World Commission on Environment and Development) (1987) *Our Common Future*, Oxford: Oxford University Press.

8 Integrating environment into regional economic-development plans through strategic environmental assessment

Elizabeth Wilson

Introduction

One measure for integrating environmental considerations into economic development which has received considerable nominal support amongst governmental and non-governmental groups is the strategic environmental assessment (or SEA) of development-related policies and programmes. This chapter examines some recent initiatives where SEA has been applied to evaluate the environmental impacts of economic development at the regional level in the UK. It reviews practical experience with SEA in the mid- to late 1990s and concludes with some reflections on the scope for SEA in the new approach to regional governance proposed in 1997 by the incoming Labour Government.

There are a number of critical issues relating to the environmental appraisal of policies and plans at the regional scale: the scope of broad economic objectives, and their translation into the promotion of economic development; the relationship between different levels of government (European, central and local government) and regional agencies; the nature of policy implementation; and the practical feasibility of strategic appraisal. A key issue is the nature of the policy-making and implementation process, where overall economic objectives adopted at national or supra-national level (such as promoting competitiveness, liberalization, deregulation or harmonization) set the framework for economic activity and for the public and private-sector programmes and development projects that give it shape. The profound changes to the key economic-infrastructure sectors of electricity, gas, water and transport in the UK over the decade of the 1990s illustrate the powerful role of the broader policy framework in initiating change.

The economic-development policies and programmes promoted by local government and regional agencies in the UK may suggest a goal-oriented activity with explicit intentions and with measurable outcomes attributable to that activity which lends itself to appraisal measures such as cost-effectiveness and efficiency. However, local economies are also influenced by the aspatial policies of national and supra-national government which may have unintended or unforeseen outcomes at the local scale. Local economies are also influenced by the processes of globalization which may not be within the influence of national or supra-national policies. Consequently, it may be difficult to attribute specific outcomes directly to economic-development policies or programmes; in addition, the objectives of different policy-making tiers of government may well be at variance.

These contextual issues are of critical importance in discussing the scope for the environmental appraisal of economic-development policies and programmes at the regional level. Consequently, this chapter examines experience with SEA as it relates to national and European Union (EU) policy-formulation as decisions taken at these levels set the context for the economic-development programmes of local authorities and regional partnerships. The chapter also critically reviews the experience with strategic appraisal of programmes of regional economic development funded by the EU and the impact of the appraisals that have taken place on the types of projects supported by the EU.

Sources of pressure

The stimulus for action to incorporate environmental considerations into economic decision-making at the strategic level has come from three sources which are derived from a greater understanding of the consequences of past investment decisions and from an acceptance of the need to move towards more sustainable forms of development.

First, the shortcomings of the system of project environmental assessment introduced in the EU (then the European Community) in 1988 have been evident. They have been the subject of the EU's own monitoring reports (CEC 1993a) as well as the concern of academic comment (Therivel *et al.* 1992; DoE/IAU 1996). While much of this concern has related to the quality and interpretation of the different national regulations, there has also been a recognition that issues related to the need for the development, for both public and private schemes, and the consideration of alternative policies, processes and locations, have not been adequately addressed in the European legislative framework. In particular, there has been no requirement or procedure to appraise the cumulative impacts of various projects that are separately assessed by project-focused Environmental Impact Assessment (EIA). Recent amendments to the directive (CEC 1997a) have only partially dealt with this problem.

Second, it has been recognized that national and supra-national government programmes of physical economic development and infrastructure provision can and do have adverse environmental consequences (e.g. DoT 1993). At the European level, it has been accepted that the environmental impacts of the EU's own economic policies, founded as they were on the ideal of a single internal market which since 1992 has allowed the free movement of goods, people and services within the EU, have in many instances been negative. The overall objective of attaining a single internal market has given rise to two policy areas: sectoral policies for areas of economic activity such as agriculture, fisheries, energy and transport; and cohesion and harmonization measures to achieve a level playing field for the operation of the single internal market.

In relation to the sectoral policies of the EU, there has been a gradual recognition, hard fought for by the environmental Non-Governmental Organizations (NGOs), of the damaging effects of the price support and other intervention mechanisms for the various economic sectors: the Common Agriculture Policy, for instance, favoured intensification and high inputs, leading to over-production and pressure on soils, water resources and biodiversity (EEA 1995). Similarly, in relation to the activities designed to promote harmonization and cohesion within the EU, the consequences of spending by the EU on reducing spatial and economic inequalities within Europe through its social and cohesion funds were often in conflict with its own environmental policies. For example, the EU's own Court of Auditors was critical of the weak monitoring mechanisms in place (Court of Auditors 1992) and the review of the Structural Funds revealed conflicts between the developments financed by the funds and environmental objectives (CEC 1993b). In response, environmental NGOs pressed the European Commission to consider adopting a procedure to ensure better prior appraisal of the spending programmes against environmental objectives (Baldock et al. 1992; RSPB 1992).

Thirdly, the widespread adoption of broad commitments towards more sustainable development arising from the Rio Summit of 1992 has prompted the search for better ways of integrating environmental concerns into decision-making at all levels and across all policy areas. The appraisal of environmental impacts at the time at which policies are being formulated has been widely advocated as one way of achieving this integration. Furthermore, recognition of the need for integration has not been confined to policy areas with direct, tangible environmental impacts. It has been acknowledged that policies with no explicit physical programme, such as those of privatization and deregulation, and elements of economic-development programmes in the 1990s, such as support for business competitiveness, can lead to undesirable environmental impacts. The UK Department of the Environment's guide *Policy Appraisal and the Environment* cites fiscal policies as one such example where a systematic *ex ante* assessment could be undertaken of the likely environmental consequences (DoE 1991).

The groups established in the UK to advise the government on the implementation of the commitments from the Rio Summit have also focused on the role of fiscal policies. The UK Panel on Sustainable Development, in its Third Report, argued that it is difficult to establish the total level of direct and indirect subsidies in the economy (with direct subsidies alone representing at least £7,300 million in 1995/6), and equally difficult to distinguish the precise environmental effects of any subsidy as their effects are widespread. Nevertheless, the panel has stated that 'there can be no doubt that some subsidies, including some with specific environmental objectives, have adverse environmental impacts' (British Government Panel on Sustainable Development 1997: 12). The panel cites an estimate of environmentally damaging subsidies at £20,000 million a year, and recommends that '[A]ny proposals for new subsidies should be subject to environmental impact assessment' (British Government Panel on Sustainable Development 1997: 13).

The role of strategic environmental assessment

These conclusions have led to arguments, put forward by both governmental and non-governmental organizations, for some form of environmental assessment at the strategic level at which policies and plans are proposed and adopted. Strategic environmental assessment has been defined as:

> the formalized, systematic and comprehensive process of evaluating the environmental impacts of a policy, plan or programme and its alternatives, the preparation of a written report on the findings, and the use of the findings in publicly-accountable decision-making.
>
> (Therivel *et al.* 1992: 19–20).

This definition, based on experience in the US and on the EU's early draft directives on SEA, is a strict one in the sense that it requires an explicit, systematic process of appraisal and public involvement in the process. There have been reservations about its feasibility in practice and about the capacity of government agencies to undertake strategic appraisal. Initially, the problems anticipated were both technical and political (Wilson 1993). There were doubts about whether methodologies for project environmental assessment could be adapted at the strategic scale when the scope of the possible impacts of a policy was so uncertain, the prediction of environmental impacts tenuous and the interactions complex. There were also concerns about the most appropriate procedures to employ, about the timing of the assessment process, about the authority needed to undertake or validate such an appraisal and about the delays it might cause to the policy-formulation process. There were also concerns about whether strategic appraisal is consistent with notions of subsidiarity and whether it implied a hierarchical nature of policy-formulation which did not reflect reality (Therivel *et al.* 1992).

Despite these concerns, in the UK the idea received broad support from statutory agencies such as English Nature (EN 1994), from local-authority associations such as the Local Government Management Board (LGMB 1993) and from NGOs such as the Royal Society for the Protection of Birds (RSPB) (Therivel *et al.* 1992) and the Council for the Protection of Rural England (CPRE) (Jacobs 1992). Moreover, it can be argued that the institutional capacity to respond did exist, at least in the sense of governmental commitments to the idea of strategic appraisal. At the national level, the UK government had expressed its support in 1990 for some form of strategic appraisal in its first environmental strategy, *This Common Inheritance* (HMG 1990); it published guidance on appraisal for civil servants (DoE 1991), and reiterated its support in its Sustainable Development Strategy (HMG 1994).

At the European level, both the Fourth Environmental Action Programme (EAP) which covered the period from 1987 to 1992 and the Fifth EAP, *Towards Sustainability*, which covers the period from 1992 to 2000 (CEC 1992) include a commitment to some form of SEA of community policies. The legislative measures adopted under these action plans have varied widely, from those setting precise standards for emissions or ambient environmental quality to, more recently, procedural measures such as those for EIA, access to environmental information and eco-auditing. The Fifth EAP was specifically aimed at a range of actors, including member-state governments, industry and local government, whilst also targeting certain economic sectors (agriculture, industry, tourism, energy and transport). As well as these vertical measures it proposed a raft of horizontal measures such as the provision of more and better information and comprehensive plans for common areas such as coasts.

The statements of principle and intent within these EAPs have been significantly strengthened by various treaties that establish the legal basis for the activities of the EU. For example, the Maastricht Treaty has at last given an explicit provision for environmental objectives to be reflected in all of the activities of the EU, with Article 130r(2) stating that 'Environmental protection requirements must be integrated into the definition and implementation of other Community policies' (CEC/CEC 1992). A declaration annexed to the treaty also noted that the commission and the member states undertake to take full account of the potential environmental impacts of policy proposals. However, while commitments in the form of statements of good intent perhaps demonstrate a willingness to respond to the calls for environmental appraisal at the strategic level, a definite response has proved more elusive. The next section looks at some of the features which have characterized the implementation of SEA in practice in the field of economic development.

The nature of the response

The following discussion will illustrate the nature of the response to the calls for the environmental appraisal of economic-development policies, plans and programmes by examining practice at two levels of government. First, the environmental appraisal of economic policies within central-government departments will be briefly considered. Second, the integration of environmental concerns into the regional development programmes supported by the EU Structural Funds will be more fully assessed and the involvement of government at local, central and EU levels will be considered. It should also be pointed out that local government itself has made substantial progress in a third area, the environmental appraisal of land use and development plans.

The greening of government

In practice, the response in the UK has been strongly shaped by the broader commitment of the Conservative government, which was in power until 1997, to deregulation, to rolling back the boundaries of the state and to its support for voluntary compliance with environmental goals. This approach gave rise to a reliance on guidance and information about best practice, such as that provided in 1991 on *Policy Appraisal and the Environment* (DoE 1991) and to exhortation to businesses, households and other stakeholders to improve their environmental performance as set out in the 1994 Sustainable Development Strategy (HMG 1994). This reliance on guidance and exhortation reflected a concomitant resistance to new regulatory mechanisms.

Despite this background, some new procedures were promised within central government: these included a commitment that each department should have a green minister responsible for overseeing its environmental activities, that departments would report on environmental progress in their annual departmental reports, and that papers proposing new policies or initiatives to Cabinet or Ministerial Committee 'should, where appropriate, cover any significant costs or benefits to the environment' (HMG 1994: 197). Nevertheless, central government's attitude was generally against the need for any formal requirement for SEA, a stance clearly shown in the UK response to the EU's draft Directive on Strategic Environmental Assessment. The government resisted proposals from the European Commission for a directive requiring formal SEA by arguing that a form of SEA was already being delivered through the activities mentioned above (Wilson 1993).

The original draft proposal of the EU in 1991 would have required an SEA of a wide range of policies and programmes at central-government level, including those for broad economic policy as well as for particular sectors of economic activity, undertaken in a publicly accountable manner by an independent authority (CEC DGXI 1991). This initial proposal was whittled away in subsequent years to a much reduced sphere of policy-making. The current proposal, put forward by the European Commission in 1996 (CEC DGXI 1996), requires that only plans and programmes adopted as part of the decision-making process as it relates to land use should be subject to an SEA. As these include energy, waste, water, industry (including minerals), tele-communications, tourism and some transport infrastructure, the scope is still significant, and many groups (see, e.g. CPRE 1997) have welcomed its adoption as an essential step to formal SEA. However, many elements of local and regional authorities' activities related to economic development, such as the programmes of funding, skills and employment training, business and innovation support, are less likely to be included.

The Greening of the Structural Funds

In contrast to the discretionary UK approach, new formal mechanisms to deliver some form of SEA in the largely spatially based programmes of the EU Structural Funds have been introduced. These funds comprise the European Regional Development Fund, the European Social Fund, the guidance element of the European Agricultural Guarantee and Guidance Fund, and the Financial

Instrument for Fisheries Guidance. Together with the Cohesion Fund, they represent over one-third of the EU's total budget (CEC 1997b). The Structural Funds are directed at achieving six objectives: (1) assisting lagging regions; (2) supporting adjustment in regions affected by industrial decline; (3) combating long-term unemployment; (4) assisting young people; (5a) supporting agricultural and fisheries restructuring; (5b) strengthening rural areas, and (6) promoting the economic development of peripheral areas. Some 85 per cent of the funds goes to meet objectives 1, 2, 5b and 6 in specific regional areas designated according to economic, labour market and demographic factors (CEC 1997b).

. Following the review of the Structural Funds in 1993, new regulations for the 1994–9 programme required that regional plans submitted for funding under these spatial objectives must include an assessment of their environmental impact (CEC 1993b). The earlier regulations governing the Structural Funds had required programmes to be consistent with existing EU policies, such as environmental protection, but they now have to be consistent with the sustainability objectives of the Fifth EAP. Some of the projects supported through the funds had been geared to environmental protection, such as waste-water-treatment plants or rehabilitation of industrial sites, but the programmes were normally appraised in terms of conventional economic-development objectives such as job creation. For all applications for funding support, the new regulations required that an environmental profile for the region be prepared, describing the existing environmental state of the region, and that the impact of strategies and plans be evaluated against the EU's commitment to sustainable development, and that competent environmental authorities be associated with the preparation of the regional plan.

In relation to the implementation of these regulations, as the administrator of the Structural Funds, the European Commission has been involved more actively in evaluating and monitoring the extent to which the requirement for an SEA has been operationalized within the various stages of the regional plan as set out in the Community Support Framework (CSF) or Single Programming Documents (SPDs) which constitute an agreement between the member states and the commission. Specific duties have therefore been imposed on the monitoring committees, the regional partnerships which administered the programmes, to undertake a form of SEA. These partnerships for the most part in the UK comprised representatives of the EU, government departments (particularly the Departments of Environment – now the Department of Environment, Transport and the Regions – and the Department of Trade and Industry), local authorities in the region, private-sector interests, training and enterprise councils and representatives of the voluntary sector (Seamark 1996a).

The performance of adopted measures

How far have these new procedures, mechanisms and best-practice guidance delivered better environmental outcomes?

Greening of government

The *Policy Appraisal and the Environment* (PAE) guidance issued in the UK in 1991 to encourage the integration of environmental concerns into central and local-government policies and programmes through appraisal at the strategic scale has met with mixed success. For its part, central government has claimed that the guidance has been acted upon to some degree. Following on from the guidance issued in 1991, the Department of the Environment issued a report *Environmental Appraisal in Government Departments* (DoE 1994) which purported to cite case studies where appraisal and evaluation had been used. However, the report revealed little direct progress with implementing the guidance in major departments of state, particularly for key economic sectors, and failed to convince those NGOs which had been promoting SEA that it was actually being applied in practice. The CPRE, for instance, in a review based on parliamentary answers, concluded that 'No government department can provide evidence that it has subjected its policy statements to an environmental appraisal' (CPRE 1996: para. 8). A later study commissioned by the government from outside consultants, on the other hand, gave a rather more optimistic interpretation: it concluded that '[i]n broad terms, Departments are conducting policy appraisals according to the spirit, if not the letter, of the approach set out in the PAE guidelines. But the evidence from the case studies is that there is still room for more comprehensive and systematic consideration of the environmental impacts of policy' (DETR 1997a: 1). In particular, economic-development initiatives do not appear to have been subject to systematic appraisal of their environmental impacts. For instance, one of the case studies in the consultants' report was of the Regional Selective Assistance Programme of the Department of Trade and Industry, the key part of the government's regional, industrial and inward-investment strategy. The report found that 'it is clear that any environmental costs incurred through assisted projects are not measured in any way' (DETR 1997a: 98).

It is clear from the report that some forms of appraisal and scrutiny – for instance, economic and financial appraisal – can and do receive political and institutional support; indeed a range of measures was put in place during the 1990s to further the objectives of the government's competitiveness and deregulation initiatives. Consequently, it can be argued that it is lack of political support for environmental appraisal, rather than technical difficulties in implementing it at the strategic level, which is critical (CPRE 1996).

Despite this lack of anticipatory assessment, some *post hoc* evaluation of policies in practice has been undertaken. The select committees of the House of Commons, for instance, have undertaken reviews of broad policy areas, as in the Environment Committee's report on the Environmental Impacts of Leisure Activities (HoC 1995), but these are not *ex ante* strategic appraisals. While they perform an extremely valuable function in monitoring the consequences of policies and lead to recommendations for better practice to which governments have to respond, they do not meet the criteria of prior, predictive or preventive assessment which are the strengths of SEA. Possibly offering more scope for such a strategic prior assessment are the reviews conducted by the Round Table on Sustainable Development, which have examined issues such as the broad impacts of the liberalization of the domestic energy markets, and freight transport (UKRTSD 1996). While having the advantage of an independent public profile, and being free to be critical of government policy, they nevertheless do not constitute a systematic appraisal of environmental impacts and their integration with policy formulation.

Environmental appraisal of Structural Funds

It might be expected that the more formal requirement for a form of SEA for EU Structural Funds would perform better in delivering different environmental outcomes than the discretionary approach applied at the national level in the UK. However, so far its integration has remained at the procedural level, with SEA being tacked on as a set of administrative steps of compliance rather than as an integral part of programme development; there has been little change in the outcomes in the form of types of projects selected for funding support. Nonetheless, the review of the EU's Fifth EAP, perhaps understandably, gave a fairly upbeat judgement: it noted that improvements to the CSFs and SPDs were 'a more visible environmental dimension in terms of integration of the environment in the development priorities of the Member States/regions, and in terms of the safeguards needed to ensure that EU environmental policy and legislation are respected during the course of implementation' (CEC DGXI 1997: 115).

However, changes in procedures may not always translate into clearly attributable environmental benefits. For example, an account of the appraisal of the Irish National Plan which received financial support from the EU for the period between 1994–9 illustrates that, while authorities may be perfectly willing to comply with formal regulatory requirements, it is difficult to establish precisely what changes have been made as a result (Bradley 1996). In reviewing the environmental profile submitted with the Irish National Plan, the commission drew attention to the need for further information and the scope for changes in implementation 'in order to avoid potential environmental impacts' (Bradley 1996: 167). Bradley argues that the CSF, and the resulting operational programmes, were much improved through the 'enhanced integration of sustainable development principles within economic development' (p. 168). However, the details of the changes, and hence of the ability of the monitoring arrangements to refuse funding to projects, under the operational programmes, remain vague.

More generally, an early survey by the RSPB of the draft SPDs for the rural regions applying for funds under Objective 5b found that the use of systematic environmental appraisal was sketchy with a lack of environmental information or any recognition of how the funds could be used to further environmental aims (RSPB 1994). Equally critical was the later review by Clement and Bachtler of the single programming documents (SPDs) for the North Sea regions (nine Objective 2 regions), which concluded that they 'generally failed to meet the criteria in the Framework Regulations' (Clement and Bachtler 1997: 13). They concluded that while baseline environmental information was provided, it was in descriptive rather than analytical terms; the interpretation of environmental impacts was often limited to *ex post* appraisals, with little examination of the scope for positive environmental gain, for example through targets for environmental improvement, use of environmental indicators and monitoring, or clear environmental criteria for project selection. They also found very limited indication of the degree or nature of involvement of environmental authorities in preparing or appraising the SPDs. Nevertheless, most commentators argue that the requirement to appraise the plans for their environmental impacts, even if it has not yet changed outcomes on the ground, has at least begun the process of changing cultural and institutional attitudes to economic development (see, e.g. Keiller 1997).

At the same time, there is some evidence that in the later SPDs, the appraisals are working through in terms of the projects grant-aided under the SPDs. In her recent study of the implementation of the environmental appraisal of the SPDs in English regions, Deborah Seamark (1996a) has argued that while it may be too early for full monitoring of the SPDs (as most of them did not become operational until 1995), the integration of full EA into the criteria for project appraisal and selection is critical. She examined the role of institutional structures, and found that a key role is played by the secretariat provided by the Regional Government Office (RGO) and the Monitoring Committee, comprising representatives of the principal partners in the programme area from the public, private and voluntary sectors.

Seamark (1996a) found that the relationship of the Monitoring Committee with the RGO is complex but that these committees both set the criteria for judging applications and select those projects which are to be grant-aided. Projects are selected against explicit criteria, but the environmental component of these is only small. She cites the example of the Greater Manchester, Lancashire and Cheshire (GMLC) Objective 2 SPD, where the points scored for environmental gain can only amount to twenty at most out of a total score (for other more typical economic-development aims such as additionality, leverage of private finance and job-creation) of 150. She argues that the advantages of the scoring system in matching projects to the specific aims or objectives of the SPD, which now include environmental ones, can be offset by the concentration on selecting environmentally beneficial projects (such as waste-water-treatment plants) rather than screening out environmentally damaging projects such as some tourism and transport schemes. Moreover, the assessment of projects against the criteria fails to consider the cumulative impacts of the projects – a common criticism of project-based environmental assessment (DoE 1996).

As far as the selection of projects is concerned, Seamark (1996a) argues that the system operated in GMLC might allow more harmful projects to be grant-aided than that of the East Midlands where a more comprehensive assessment of

Table 8 Predicted impacts of measures in East Midlands Objective 2 SPD (1994–6) compared to predicted impacts of approved projects under those measures

Measure	Principal predicted impacts in SPD	Predicted impacts of approved projects
1.1 Assistance for innovation	Neutral	No projects approved
1.2 Strengthening support for innovation	+: Encouraging use of clean technology will reduce pollution and increase energy efficiency	1 +: Clean technology → less pollution 2 neutral
1.3 Development and application of clean technologies	+: Clean technology will reduce pollution and increase energy efficiency	1 +: Environmental auditing → energy efficiency and environmentally responsible techniques 2 +: Supporting clean technologies 3 +: Enabling environmental audits
1.4 Training in advanced technologies	+: Promoting economic growth by developing skills poses threat to environment	No details
1.5 Telecommunications projects	Neutral	1 Neutral 2 Neutral 3 Neutral 4 Neutral 5 Neutral
2.1 Promoting SME start-up, growth and development	–: Promoting economic development poses threat to environment. 'It will be important to see that principles of sustainable development are encouraged'	No projects approved
2.2 Improving access to capital	–: As 2.1	1 +: Providing footpath trail linking tourism attractions
2.3 Developing the tourism industry	–: Encouraging increase in tourism could lead to more car journeys and add to visitor pressures +: Positive impacts from improved access, image and landscape. Wider range of attractions may disperse visitor pressures	
2.4 Training for business needs	–: Strengthening skills will promote economic growth and pose threat to environment +: Training for clean technologies may have positive impacts	No details
3.1 Support for diversification of activities in traditional sector companies	+: New processes expected to be less polluting than traditional industries	1 Neutral 2 Neutral 3 Neutral

		4 Neutral
		5 Neutral
		6 Neutral
		7 Neutral
3.2 Increasing inward investment	Neutral	1 Neutral
3.3 Developing sites and premises	+: Cleaning-up of brownfield sites will have positive impacts	1 +: Cleaning brownfield site
		2 +: Restoring farm buildings for training places. Improvement to SSSI and ancient monuments
	−: Risk of increase in number of motor trips	3 +: Restoration of old mill to provide workspace
		4 +: Restoration of old colliery site for workspace
		5 +: Incorporating business-support facilities whilst maintaining historic character of buildings
		6 Neutral
3.4 Improving transport infra-structure directly linked to economic development	−: Risk of adverse effects from promoting economic development	1 +: Encouraging use of public transport
	+: Positive effects from reducing congestion and increasing use of public transport	2 +: Encouraging use of public transport
		3 +: Encouraging use of public transport
		4 +: Encouraging use of public transport
3.5 Developing human resources	+: Improving local facilities should decrease trip lengths	No details
	+: Use of derelict buildings will improve urban landscape	
4.1 Environmental improvements	+: Restoration of derelict sites will improve landscape and reduce need for greenfield sites for develop-ment	1 +: Improving urban environment, building footpaths indirect
		+: improving local access to training reducing trip length

(continued)

Table 8 (cont.)

Measure	Principal predicted impacts in SPD	Predicted impacts of approved projects
		2 Neutral
		3 +: Landscape and urban enhancement
		4 +: Demolishing derelict buildings, preparing brownfield site for development
		5 +: Improving historic buildings, building footpaths
		6 +: Developing brownfield site
		7 +: Restoring derelict buildings
		8 +: Landscaping and wildlife protection
4.2 Developing economic opportunities	Neutral	1 Neutral
4.3 Improving access to training and employment	−: Risk of negative impacts from promoting economic growth +: Positive effects from reducing congestion and increasing use of public transport	No projects approved
4.4 Developing pathways to employment	+: Improving local facilities will mean fewer car journeys +: If derelict buildings are used, will improve urban landscape	1 Neutral

Source: RGO – East Midlands, 1996, as in Seamark (1996a).

the positive and negative effects of projects is undertaken. Table 8 shows the criteria used in the East Midlands SPD. Her overall conclusion is that environmental appraisal of regional plans could be used much more effectively as a tool for delivering more sustainable forms of development (Seamark 1996b).

Scope for future progress

What these examples illustrate is the development during the 1990s of a range of measures, from regulation to exhortation, that seek to integrate environmental appraisal into the decision-making process as it relates to the formulation of economic-development programmes and plans. The examples also illustrate the varied efforts of agencies at different levels of government to take up the initiative. As far as SEA is concerned, the experience gained in the 1990s has shown that it has not been technical difficulties, as predicted earlier at a time when knowledge and experience were patchy (see, e.g. Therivel *et al.* 1992; Wilson 1993; Burleigh 1993) that have hindered progress. On the contrary, there has been a considerable advance in practice both procedurally and methodologically at the level of programmes and plans (Therivel and Partidario 1996). The feasibility of SEA is therefore established at least to some degree.

However, the examples discussed above demonstrate the difficulties of devising a system of strategic environmental appraisal which clearly and accountably brings about different outcomes in economic-development decisions on the ground. The reasons for this lie not just in the complexity of dealing with alternatives and impacts at this level, and in the lack of institutional structures to implement good intentions, but also in the lack of political will. The weaknesses of policy documents with good intentions but without the necessary powers of implementation in the face of reluctant vested economic interests and restricted institutional competencies are revealed at supra-national and local level. For instance, the interim review of the EU's Fifth EAP concluded that '[A]ctions being taken to date will not lead to full integration of environmental considerations into economic sectors or to sustainable development' (EEA 1996: 1). In particular, the interim review argues that there has been little change in the underlying production and consumption trends, with further growth forecast in the target economic sectors, especially those of tourism and transport.

In the absence of major structural economic change, it is possible to suggest that there are three ways in which the culture of economic development can respond further to the environmental cause through SEA. These proposals build on the growing body of experience with the SEA of plans and programmes.

Monitoring and evaluation

One relatively straightforward option is for more attention to be paid to the effectiveness and efficiency of measures already adopted. The final report on the EU's Fifth EAP commented that, '[W]hilst there have been improvements in ensuring better integration of environmental considerations into the use of the Community's financial support mechanisms, there is a continuing need to improve the evaluation of the impact of such funding to avoid unsustainable approaches' (CEC DGXI 1997: 11). It points out that most expenditure under the funds has fallen within those sectors which the Fifth Action Programme warned might have negative impacts, and suggests three areas for consideration to improve further the implementation of the programmes: 'to monitor how much of the CSFs and SPDs in Objective 1, 2 and 5b areas was in both the preventative and curative environmental field; to look at the responsibilities of the designated environmental authorities; and to enhance the environmental dimension of existing programme monitoring and environmental procedures' (p. 115). These points appear to support the conclusions of Clement and Bachtler (1997) that none of the SPDs they reviewed integrated the environment effectively. Clement and Bachtler nevertheless remained optimistic that the provisions of the regulations could be improved, particularly in the areas of data provision, environmental integration, environmental targets and environmental gain. Both they and Seamark (1996a) suggest that the agencies involved, in particular member-state governments and the commission, could contribute more by giving guidance on methodologies for better integration and for the appraisals themselves.

This need for guidance is evident at the national level, at least in the UK, where there is a need for more targeted guidance on environmental appraisal in government departments (DETR 1997a) and for better take-up of existing guidance. This is an area where central government could learn from the more innovative practice at regional and local level. Going beyond improved implementation of existing guidance, in the UK new institutions such as the Environmental Audit Committee of the House of Commons could potentially develop a role in strategic policy evaluation rather than *post hoc* auditing.

Sustainability appraisals

While the EU still deliberates on whether to adopt the proposed SEA directive even in its currently limited form, there has been an interesting extension of the scope of strategic environmental appraisals: a number of local authorities in the UK, in reviewing their land-use-development plans, are extending what were initially environmental appraisals into sustainability appraisals. This involves a broader assessment against not just physical environmental criteria but also socio-economic criteria such as number or type of jobs, and more intangible measures of quality of life such as sense of ownership and participation, community vitality, accessibility, literacy and numeracy, and sense of security. Sustainability appraisals are likely to include a policy consistency check, as in the first step in environmental appraisals, and so offer authorities the chance to integrate the objectives of their economic-development plans with, for instance, their land-use plans, transport plans and programmes, housing programmes and tourism strategies.

In the past, local authorities' activities in the sphere of economic development consisted of promoting business activity, employment and wealth creation in an area for the benefit of the local community. This has usually been the responsibility either of a separate department or of one attached to planning or the Chief Executive's Office. While the professional base may be partly drawn from planning, units often have an explicit remit to develop links with the business community and with training and enterprise agencies. Consequently, they may become isolated from broader planning concerns. This is significant because some of those who have commented on the substantial changes to the powers and process of planning in the 1980s and 1990s (e.g. Brindley *et al.* 1996; Wilson 1998) have argued that land-use planning has to some extent revived its declining fortunes and intellectual energy by embracing the environmental issue. Thus, the separation of some aspects of economic development from planning may have an adverse effect. Sustainability appraisal may provide an opportunity for economic-development teams to redefine their roles in relation to environmental objectives and authority-wide quality-of-life commitments.

This change in the culture of economic development, and a reinterpretation of the usual quality-of-life measures, could promote better integration of environmental and economic strategies: not only through using environmental appraisal criteria, and more explicit environmental objectives, but by greening local economies through promoting certain projects. The process of sustainability appraisal could lead to an adjustment of economic-development programmes to deliver environmental projects such as biodiversity conservation, environmental management systems, or more environmentally friendly transport modes.

These broader appraisals may be further stimulated by change in the provision of support for economic development brought about by financial stringency. The current review of EU regional policy and of the structural funds suggests that the form of economic development supported will change from that of major infrastructure projects towards more use of financial, innovation and training support (Healey 1997). In institutional terms, this may promote greater integration of these programmes with those of the regional and local-government agencies and the partnerships responsible for delivering them.

Innovation in governance

The 1997 white paper proposing the establishment of regional development agencies in England makes sustainable development one of the five objectives for the agencies (DETR 1997b) (see Table 9). This presents another opportunity to integrate economic-development objectives with environmental concerns. While the prime function of the agencies will be to prepare and implement a regional

Table 9 Objectives for regional development agencies

1 Economic development and social and physical regeneration
2 Business support, investment and competitiveness
3 Enhancing skills
4 Promoting employment
5 Sustainable development

Source: DETR (1997b: 17).

economic-development strategy, the document proposes that they be given a statutory objective to further sustainable development. It argues that social and economic development and environmental conservation need integrating to achieve this, and that 'underperformance in an economy risks putting a greater pressure on the environment' (DETR 1997b: 39). The agencies will be expected to further environmental concerns by promoting resource efficiency (such as waste minimization and energy conservation) and to take a lead on the EU Structural Funds and their successor programmes.

How the relationship between the regional development authorities and democratically elected tiers of local government, business interests, NGOs and statutory environmental agencies will develop is unclear. While the agencies will have a 'clear, predetermined level of local government on their boards' (DETR 1997b: 49), they will be explicitly 'business-led' (p. 16). Neverthelesss, the potential exists for a statutory obligation to prior, strategic sustainability appraisal of the agencies' programmes, which would be consistent with the revised regional-planning-guidance documents which are themselves to be subject to sustainability appraisal.

Conclusions

The conditions for the wider and more successful application of SEA can be met. These conditions relate to the degree of political support, the procedures to undertake it and the evidence of its technical feasibility. This chapter has tried to show that in the UK local government and regional agencies have begun to develop good practice in this area. However, some difficulties perhaps naturally remain. In particular, the rational and sequential conception of policy formulation which SEA implies relies on an unrealistically hierarchical form of decision-making that may not be reflected in reality. Significantly, it could be that the culture of economic development is still resistant to the broader integration of environmental concerns. More prosaically, it could be that it takes time and experience to adjust well-established modes of operation and ways of thinking in order to exploit to the full the new procedures, instruments and institutions that have the potential to integrate the concerns of a wide spectrum of communities and stakeholders into the key economic decisions which affect them.

References

Baldock, D., Beaufoy, G., Haigh, N., Hewett, J., Wilkinson, D. and Wenning, M. (1992) *The Integration of Environmental Protection into the Definition and Implementation of Other EC Policies*, London: IEEP.

Bradley, F. (1996) in R. Therivel and M. Partidario (eds) *The Practice of Strategic Environmental Assessment*, London: Earthscan, pp. 157–68.

Brindley, T., Rydin, Y. and Stoker, G. (1996) *Remaking Planning: the Politics of Urban Change*, second edition, London: Routledge.

British Government Panel on Sustainable Development (1997) *Third Report*, London: Department of the Environment.

Burleigh, D. (1993) 'Integrating environmental concerns into European Community regional policy', unpublished MSc dissertation, Oxford Brookes University, Oxford.

Clement, K. and Bachtler, J. (1997) 'Regional development and environmental gain: strategic assessment in the EU Structural Funds', *European Environment* 7, 7–15.

CEC (Commission of the European Communities) (1992) 'Towards sustainability: A European Community programme of policy and action in relation to the environment and sustainable development', COM(92)23 Final, Brussels: EC.

CEC (Commission of the European Communities) (1993a) *Report from the Commission of the Implementation of Directive 85/337/EEC on the Assessment of the Effects of Certain Public and Private Projects on the Environment*, COM(93)28 Final, vol. 12, Luxembourg: Office for Official Publications of the European Communities.

CEC (Commission of the European Communities) (1993b) *Community Structural Funds 1994–1999: Revised Regulations and Comments*, Luxembourg: Office for Official Publications of the European Communities.

CEC (Commission of the European Communities) (1997a) *Directive 97/11/EC Amending Directive 85/337/EEC on the Assessment of the Effects of Certain Public and Private Projects on the Environment*, OS L73.

CEC (Commission of the European Communities) (1997b) *Agenda 21: the First Five Years. European Community Progress on the Implementation of Agenda 21 1992–1997*, Luxembourg: Office for Official Publications of the European Communities.

CEC DGXI (Commission of the European Communities DGXI) (1991) *Draft Proposal for a Directive on the Environmental Assessment of Policies, Plans and Programmes*, XI/194/90-EN-REV-4, Brussels: CEC.

CEC DGXI (Commission of the European Communities DGXI) (1996) *Proposal for a Council Directive on the Assessment of the Effects of Certain Plans and Programmes on the Environment*, Brussels: CEC.

CEC DGXI (Commission of the European Communities DGXI) (1997) *Towards Sustainability: The European Commission's Progress Report and Action Plan on the Fifth Programme of Policy and Action in Relation to the Environment and Sustainable Development*, Luxembourg: Office for Official Publications of the European Communities.

CPRE (Council for the Protection of Rural England) (1996) *Greening Government: From Rhetoric to Reality*, London: CPRE.

CPRE (Council for the Protection of Rural England) (1997) *Proposed Directive on Strategic Environmental Assessment: A Campaign Briefing Pack*, London: CPRE.

CEC/CEC (Council of the European Communities/Commission of the European Communities) (1992) *Treaty on European Union* (The Maastricht Treaty), Luxembourg: Office for Official Publications of the European Communities.

Court of Auditors (1992) 'Special Report no. 3/92 concerning the environment together with the commission's replies', *Official Journal of the European Communities*, C245.

DoE (Department of the Environment) (1991) *Policy Appraisal and the Environment*, London: HMSO.

DoE (Department of the Environment) (1994) *Environmental Appraisal in Government Departments*, London: HMSO.

DoE/IAU (Department of the Environment/Impact Assessment Unit) (1996) *Changes in the Quality of Environmental Statements for Planning Projects*, London: HMSO.

DETR (Department of the Environment, Transport and the Regions) (1997a) *Experience with the 'Policy Appraisal and the Environment' Initiative*, London: DETR.

DETR (Department of the Environment, Transport and the Regions) (1997b) *Building Partnerships for Prosperity: Sustainable Growth, Competitiveness and Employment in the English Regions*, London: DETR.

DoT (Department of Transport) (1993) *The Government's Response to the SACTRA Report on Assessing the Environmental Impact of Road Schemes*, London: HMSO.

EN (English Nature) (1994) *Sustainability in Practice*, Peterborough: EN.

EEA (European Environment Agency) (1995) *Europe's Environment: the Dobris Assessment*, Copenhagen: EEA.

EEA (European Environment Agency) (1996) *Environment in the European Union: Report for the Review of the Fifth Environmental Action Programme*, Luxembourg: Office for Official Publications of the European Communities.

Healey, A. (1997) 'Developing an interest in Europe', *Planning*, 14 March, p. 13.

HMG (Her Majesty's Government) (1990) *This Common Inheritance*, Cmnd 1200, London: HMSO.

HMG (Her Majesty's Government) (1994) *Sustainable Development: The UK Strategy*, Cm 2426, London: HMSO.

HoC (House of Commons Select Committee on the Environment) (1995) *The Environmental Impacts of Leisure Activities*, London: HMSO.

Jacobs, M. (1992) *Sense and Sustainability*, London: CPRE.

Keiller, A. (1997) 'Strategic environmental assessment in Scotland', *EA: the Magazine of the Institute of Environmental Assessment and the Environmental Auditors and Reviewers Association*, September, pp. 26–7.

LGMB (Local Government Management Board) (1993) *A Framework for Local Sustainability: A Response by UK Local Government to Agenda 21*, Luton: LGMB.

RSPB (Royal Society for the Protection of Birds) (1992) *Greening Europe: The RSPB Environmental Vision for the European Community*, Sandy: RSPB.

RSPB (Royal Society for the Protection of Birds) (1994) *Draft Single Programming Documents for Objective 5b in the United Kingdom*, Sandy: RSPB.

Seamark, D. (1996a) 'Structural funds and sustainable development: Environmental appraisal of single programming documents in the English regions', unpublished MSc dissertation, Oxford Brookes University, Oxford.

Seamark, D. (1996b) 'European funding and environmental appraisal', *Town and Country Planning* 65(12), 340–1.

Therivel, R., Wilson, E., Heaney, D., Thompson, S. and Pritchard, D. (1992) *Strategic Environmental Assessment*, London: Earthscan.

Therivel, R. and Partidario, M. (eds) (1996) *The Practice of Strategic Environmental Assessment*, London: Earthscan.

UKRTSD (United Kingdom Round Table on Sustainable Development) (1996) *First Annual Report*, London: Department of the Environment.

Wilson, E. (1993) 'Strategic environment assessment: Evaluating the impacts of European policies, plans and programmes', *European Environment* 3(2), 2–6.

Wilson, E. (1998) 'Planning and environmentalism in the 1990s', in H. Thomas and P. Allmendinger (eds) *Urban Planning and the British New Right*, London: Routledge.

Part IV
Regional and local case studies

9 Integrating environment into regional plans and strategies

The case of the West Midlands

Chris Carter and Steve Winterflood

Introduction

Central government in the UK has generally perceived the town and country planning system as a means of regulating the development and use of land in the public interest. In particular, development plans are seen as the most effective way of striking the right balance between the demand for development and the protection of the environment. It is the plan-led planning system, which was given statutory force in the 1990 Town and Country Planning Act (introduced through the Planning and Compensation Act, 1991), which has dominated the development of town and country planning in the UK throughout the 1990s.

Within the context of the national planning framework, local planning authorities have been given new freedoms to influence the nature of future development within their areas via the development-plan system. However, these new freedoms have been accompanied by an unprecedented growth in the development of national and regional planning guidance. The 1990 Town and Country Planning Act requires local planning authorities to have regard to this guidance in preparing their development plans. If local planning authorities fail to take this guidance into account, the Secretary of State reserves the right to intervene.

This chapter seeks to explain and evaluate how the local authorities in the West Midlands prepared advice for the national government on the type of regional planning guidance (RPG) they would wish to be issued for the West Midlands. The process of preparing this advice had at its centre the challenge of integrating proposals for economic growth with the aim of working towards environmental sustainability.

The West Midlands Region

The West Midlands Region occupies a central position in England. The population of over 5 million people live in the seven metropolitan districts of the West Midlands conurbation and in the four shire counties of Hereford and Worcester, Shropshire, Staffordshire and Warwickshire.

The region is one of great diversity. At its heart is the West Midlands conurbation which consists of Birmingham, the UK's second-largest city, the Black Country, Solihull and Coventry. This metropolitan area is densely populated, consisting of many deprived districts and containing a legacy of industrial disinvestment. It also contains a range of regional centres and activities including Birmingham City Centre, five universities, cultural venues of international importance, unrivalled sporting facilities and, of course, the manufacturing powerhouse of the region.

Surrounding the West Midlands conurbation and separated from it by the green belt is a ring of smaller towns and extensive rural areas of high landscape and environmental quality. These areas have received rapid population growth in recent years associated with post-war decentralization from the conurbation. These towns also have significant links to the conurbation for work, social and leisure purposes. In the north of the region is another conurbation, that of North Staffordshire, which is a relatively self-contained urban area whose economy is based traditionally on the ceramics industry and endures on a smaller scale many of the problems of the West Midlands conurbation.

The majority of the region is rural, 70 per cent of the land is used for agriculture, which has continued to suffer a significant decline in employment. The region has a rich and varied landscape and townscape. It includes some of the UK's finest hill country boasting several areas of outstanding natural beauty and many historic towns including Stratford-on-Avon, Worcester, Lichfield and Hereford. There are also a number of important nature-conservation areas including the Ramsar sites of the Midland Reserves and Mosses.

The West Midlands Regional Forum of Local Authorities

This forum is an association of the county and shire districts of Hereford and Worcester, Shropshire, Staffordshire and Warwickshire, the metropolitan district councils of Birmingham, Coventry, Dudley, Sandwell, Solihull, Walsall and Wolverhampton, and CENTRO (the corporate name of the West Midlands Passenger Transport Executive).

The forum provides a mechanism, determined by member authorities, to further their collective interests, which enables them to:

- work together and, with their regional partners, to enhance the quality of life of the inhabitants of the West Midlands Region as a whole and to improve and sustain its economy and environment in ways they cannot do individually;
- receive, analyse and disseminate information about the region as background to policy-formulation locally and regionally;
- lobby collectively in the interests of the region within and beyond its boundaries in the UK and Europe;
- take such other initiatives as may be agreed from time to time that are necessary to secure the common aims of the member authorities.

In response to these objectives and to the advice in Planning Policy Guidance Note 12 (PPG12), which stated that the county councils and organizations representing other local authorities will generally have a leading role in considering and advising on regional planning matters, the West Midlands Forum in 1991 commenced the preparation of its advice to the Secretary of State for the Environment regarding regional planning guidance (RPG) for the West Midlands.

The primary function of RPG is to provide a framework for the preparation of structure plans and unitary development plans taking into account national planning policies. PPG12 notes that RPG should generally cover those issues which are of regional importance or which need to be considered on a wider geographical basis than that of local development plans. Topics covered depend on the individual circumstances of each region, but should cover priorities for the environment, transport, infrastructure, economic development, agriculture, minerals, waste treatment and disposal.

Preparation of advice

The process of preparing advice to the Secretary of State on the nature of the RPG that will guide the development and regeneration of the West Midlands up to 2011 began with the preparation of a report which sought to identify the appropriate questions that needed to be answered. This report provided an initial overview of the region, its economy and well-being. It also reviewed the strategic policies already in existence in the region and the key questions that needed to be raised. The key questions asked were divided into five interrelated areas:

- a sustainable environment – What is it that we wish to conserve or improve? What is the impact of current policies having on the region's environmental assets? What needs to be done?
- economic regeneration and restructuring – What are the choices for regional economic growth? How far can the benefits of economic growth be targeted to those areas and groups in greatest need? How can the proposals for economic growth and enhanced image be achieved?
- housing needs and requirements – What are the future housing requirements of the region? What resources will these requirements need and how will further land releases affect investment in the existing housing stock? What should be the geographical distribution of new housing?
- accessibility and mobility – What will be the transport needs of the region's residents and economy? How will these best be met and what should be the balance of provision between public and private transport? What are the implications of minimizing the need to travel so as to reduce energy consumption and atmospheric pollution? What opportunities are there to secure additional resources for transport infrastructure in the region?

- the physical structure of the region – What patterns of development will accommodate sustainable economic growth within a sustainable environment?

A number of technical groups were then set up by the forum to address these questions. The main findings of the technical work undertaken by these technical groups (of which one was the Sustainable Environment Working Group) were presented in a second report. This report introduced the issues that needed to be faced and the opportunities that could and should be taken, the general strategic aims in managing change in the region and the strategic choices that needed to be made. A consultative conference was then held to discuss the areas where common agreement was not found. The main areas of discussion related to housing, transport, the economy and the environment. On the basis of the preceding work, the final stage in the process was to prepare the advice to be submitted to the Secretary of State on the type of RPG the West Midlands Forum would wish to see issued by the Secretary of State. This advice was submitted to the Secretary of State in October 1993.

Environmental management – strategic policy prior to RPG advice

No single source of authoritative regional planning policy existed prior to the preparation of Regional Planning Guidance for the West Midlands. However, guidance had been issued for the sub-regional level through PPG10 'Strategic Guidance for the West Midlands', the key objectives of which included 'revitalising the sub-regional economy and regenerating the inner city areas'. These two key objectives were encapsulated in the aims of the advice. There was, however, one issue which was becoming of increasing and prominent importance in the preparation of the advice. This was the concern for the environment. The importance of moving to a more desirable and sustainable environment was not seen to be only an aim in itself, but also an aim which should pervade and influence other regional objectives.

The RPG advice process accepted that environmental considerations were a key part of all decision-making. It was agreed by the forum that the environmental perspective must be focal and that it should not be seen as subsidiary to economic and social concerns. Even in areas that faced severe economic problems there were considered to be good arguments for such an approach. A high-quality environment was seen as vital in helping to attract inward investment and the creation of new jobs. Advice was to have at its centre therefore the aim of achieving a sustainable regional environment. The aim of working towards environmental sustainability was the vital ingredient underpinning the whole of the advice and represented the major new force in regional strategic planning for the West Midlands.

Integrating environmental concerns

To assist in identifying the environmental implications of the advice, and to ensure that environmental concerns were fully considered in an integrated way, the forum formed the Sustainable Environment Working Group (SEWG). Mindful that many of the local authorities had a limited amount of knowledge and experience in environmental matters, membership of the SEWG was extended to include officers with specialist environmental knowledge. Therefore, in addition to the shire and district local-authority officers, the SEWG included officers from the Countryside Commission, English Nature, the National Rivers Authority (subsequently the Environment Agency), the Ministry of Agriculture, Fisheries and Food, the Forestry Authority and the Regional Sports Council. Officers from Dudley Metropolitan Borough Council and Hereford and Worcester County Council accepted the role of co-chair for the group.

In 'building the environmental window' through which all proposals had to pass, it was first necessary to identify those resources that were to be conserved or improved and to establish how these resources be best managed. The technical work of the SEWG addressed this question and identified the important natural and built resources within the region, the threats posed to them and consequently their vulnerability. The SEWG also identified the opportunities for action which would allow environmental sustainability to be promoted. Without the provision of this baseline knowledge on the region's environmental assets, it would not have been possible to understand and evaluate the impact of current and proposed policies.

The results of this work highlighted the fact that the region contained a wealth of resources some of which were assets in their own right whilst others represented an opportunity to improve the regional environment; for example derelict land. The region's resources were considered under fourteen headings: air, water, attractive and historic countryside, valuable flora and fauna and their habitats, agriculture, woodland, archaeology, amenity areas, historic and attractive townscapes, existing settlements, culture, aggregates, minerals and waste, and dereliction.

Assessing the baselines and impacts

It was acknowledged that the resources identified within the region which needed to be protected were all vulnerable to influences which could lead to the denudation of their value or to the loss of the resource altogether. The work of the SEWG therefore concentrated on identifying the likely positive and negative effects of the various influences on these resources and to identify ways of addressing the negative effects. The influences identified were both those that could be changed through the development-plan process and hence could be influenced by regional planning advice and those which could not be changed as part of this process but that nevertheless still warranted attention. These influences were set out in a matrix displaying both the negative and positive effects of the twenty-one influences identified in order to ascertain the vulnerability of the region's resources to particular types of action. It should be noted that SEWG only identified the direction of the trend and not its magnitude. The matrix (see Table 10) demonstrated which influences had a positive effect (+) and which had a negative effect (−) and those which could be positive or negative under different circumstances (✱). Those where sufficient evidence was not available to provide a clear judgement are marked with a ?.

The influences that were deemed to be susceptible to the influence of the development plan were climatic change, built-development pressures, recreation and tourism-development pressures, woodland planting, costs and availability of energy, minerals aggregate, land cost and availability, green belt and other areas of restraint, mobility and accessibility, vehicle emissions and pollution. In addition to the threats for each of the region's resources, opportunities for action were identified which could help to maintain their sustainability and allow improvements. These opportunities were identified and used in the formulation of the environment-led strategy.

The matrix therefore provided an assessment of the region's resources and the influences affecting them both in terms of threats and opportunities for action. Having undertaken this baseline assessment, questions remained about whether it would prove to be an effective environmental management tool and, if so, how it would help in the preparation of regional planning advice.

The first test was to assess the impact of current policy. To do this two further matrices were developed. A second matrix (see Table 11) was prepared to ascertain whether current policy increased or decreased the influences identified in Table 10. For example, the provision of premium industrial sites at motorway junctions obviously has an increasing influence on vehicle emissions in view of the fact that the locations encourage travel to work by car. A further example is that the use of existing buildings and land decreases the influence of built-development pressure. A third matrix was then prepared to assess whether the current regional policies have a positive or negative impact on the region's resources. The use of these matrices allowed the impact of current policies on the region's environmental resources to be assessed in a logical and consistent way.

Each regional policy was assessed and the following provide examples of the conclusions reached on the impact on the region's environmental resources of what were then the current regional economic policies:

- policies relating to the provision of peripheral premium industrial sites had an adverse effect due to direct damage to resources and through the development of green space;
- policies for out-of-town retailing attract investment away from town centres and thereby contribute significantly to air pollution through the creation of increased car-borne trips and work against the region's environmental sustainability;
- policies for concentrating development within existing settlements help to

maintain the existing building stock, resources, land and buildings whilst also helping to reduce waste generation and promote the recycling of waste and materials. As all of these factors help to reduce activities which cause climatic change and pollution they help to sustain global, national and regional environmental resources.

From this exercise, policies that were consistent and inconsistent with the goal of sustaining the environment were identified. These policy relationships are summarized in Table 12.

There are also policy areas that do not fall within the influence of regional land-use planning which have positive and negative impacts (e.g. agricultural practices) and these must also be borne in mind.

Developing the regional environment-led strategy

On the basis of the assessments conducted above, the first question asked was 'Is the current protection afforded to the region's environment adequate to meet the needs and pressures for change facing the region?' A second question then followed, namely 'If the current level of protection is not sufficient what policy changes are needed to promote sustainability and to improve the regional quality of life?'.

The matrices introduced (Tables 10 and 11) were used to assess the impacts of current policies to inform debate on the first question. The matrices were also used to evaluate the options and strategic choices put forward by the SEWG and other working groups. The SEWG used this environmental management technique to formulate a view on how the region could best move forward to 2011 by developing an environmental-led strategy for development within the region.

The environmental management technique, whilst allowing options emanating from other topic groups to be tested, considered the environment-led strategy to be the best way forward in terms of environmental sustainable policies for the region. The only option related to the pace at which the region should move towards a sustainable environment.

Strategy introduction

The move towards a sustainable environment is synonymous with the pleasant living and working conditions associated with a good quality of life. It would also introduce the conditions required to regenerate the conurbations and create the improved image which is necessary to revitalize the economy. Clean air and water supply, a balance between the built environment and green space, the presence of healthy natural features like trees and wildlife, a sense of history, easily accessed open space, shops, schools and a pleasant place of work as well as place to live, should be common to any strategy. Conversely, strategies led by economic or even social concerns that do not address environment concerns are likely to perpetuate many of the region's problems.

Table 10 Assets and influences of the West Midlands Region

Influence column key:

- A Climatic change
- B Built-development pressures
- C Recreation and tourism development pressures
- D Woodland planting
- E Costs and availability of energy
- F Mineral/aggregates (location and availability)
- G Land cost and availability
- H Green belt and other areas of restraint
- J Mobility/accessibility
- K Vehicle emissions
- L Pollution
- M
- N Intensive agriculture
- O Set aside
- P Diversification
- Q Financial (including levels of investment, grants and subsidies)
- R Water resources
- S Resource quality
- T Agricultural land quality
- U Public demand (including environmental education)
- V Technical innovation (including recycling, etc.)
- W Management of wildlife sites
- X

Columns A–M: Influences susceptible to development plans. Columns N–X: Other influences.

ASSETS		INFLUENCES	A	B	C	D	E	F	G	H	J	K	L	M	N	O	P	Q	R	S	T	U	V	W	X
Townscapes	Historic/attractive buildings/urban form/layout/spaces	1	−	−	*	+			−	*	*	−	−					?	−	+		+	+		
	Unattractive buildings/urban form/layout/spaces	2	−	+	*	+	+		+	+	−	−	−					?	−	−		+	+		
Existing settlements	Historical investment	3	−	*	*	+	+			+	*	−	−				+	?	−	+		+	+		
	Housing and infrastructure availability	4	−	*	*	+	*	+	*	+	*	−	−				+	?	−	+	+	−	+		
	Service provision	5	−	*	+		*	+	*	+	*		−			?		?	−	+		−			
	Employment provision	6		−	+	+	*		*	+	*	−	−			−	+	?	−	+			+		
	Proximity of employment and residences	7		−	+	+	*	+	−	+	−	−	−				+	?	*	*		−	+		
Amenity areas in and around settlements	Green belt/urban fringes	8	−	−	+	+	*	+	−	+	−				*	−	*	?	+	+	+	+			
	Linear open spaces/green wedges	9	−	−	+	+	*		−	+	−	−			*		*	?	+	+	+	+			
	Country parks...	10			−	*	*			−	*				−		−	?	−			−			

	Air as agent for pollution dispersal	13																							
Water	Water from rivers	14	?	✱	✱	−	+	−			−	−	−		−	+		?		✱		−	+		
	Water from reservoirs	15	?	✱	✱	−	+				−	?	−		−	+		?		✱		−	+		
	Water from aquifers	16	?			✱		?		+	−	?	−		−	+		?		✱		−	+		
	River systems and canals as agents for drainage	17	?	−		−			−		−	−	−					?					+		
	River systems and canals as agent for effluent disposal	18	?	−		−		?			−	?	−		−	+		?					+		
	Flood plains and flood defences	19	−	−		−		?		+					−			?	−				+		
	Catchments, yields and areas	20	✱	−		−		−		+			−		−			?	−						
	Recreation resource	21			−	+		+			✱		−		−		+	?	−	✱					
Countryside	Areas of landscape value	22	−	−	✱	+	+	−	−	+	✱	−	−		−	✱	✱	?	+	✱	+	+	+		
	Landscape features	23	−	−	✱	✱	+	−	−	+	✱		−		−	✱	✱	?	+	✱		+	+		
	Unattractive landscapes	24	−	✱	✱	−		+	?		−		−		−	✱	✱	?	+	−	−	+			
Derelict land	Potential/actual open space	25	+	+	+	?	+	+	✱	+	?	?	−					?		✱		+	+		
	Potential development land	26		+	+	+	+		✱	+	?	?	−					?		✱		✱	+		
Minerals and waste disposal	Mineral resources	27		−	−	+	+	−		−					?				−	✱	−	−	+		
	Aggregate resources	28		−	−	+		−		−					?				−	✱	−	−	+		
	Landfill site capacity	29		−	+		+		−				−					?							
	Potential landfill sites	30			+	+	+	+		−								?	−		−	−	+		
	Incineration facilities	31		+		+	+		+	✱			−					?					+		
	Recycling of waste	32		+			+	+	+	+			+					?		✱		+	+		
	Waste resource – for energy generation	33		+			+											?		✱		+	+		
Woodlands	Woodlands – historic and nature conservation value	34	?	−	−	✱	+	−		+	✱	?	−		−	+	✱	?	✱	✱		+	✱		
	Woodlands of timber production value	35	?	−		+	+		−	−	+	?	−			+	+	?		✱	−	−	✱		
	Woodlands of recreation landscape importance	36	?	−	−	+	−	−	−	+	+	?	−		−	+	✱	?	✱	✱		+	✱		
	Hedgerow trees and individual trees	37	?	✱	✱	−	−	−		+		?	−		−	+	✱	?	✱	✱		+			

(continued)

Table 10 (cont.)

			A	B	C	D	E	F	G	H	J	K	L	M	N	O	P	Q	R	S	T	U	V	W	X
INFLUENCES			Climatic change	Built-development pressures	Recreation and tourism development pressures	Woodland planting	Costs and availability of energy	Mineral/aggregates (location and availability)	Land cost and availability	Green belt and other areas of restraint	Mobility/accessibility	Vehicle emissions	Pollution		Intensive agriculture	Set aside	Diversification	Financial (including levels of investment, grants and subsidies)	Water resources	Resource quality	Agricultural land quality	Public demand (including environmental education)	Technical innovation (including recycling, etc.)	Management of wildlife sites	
ASSETS			Influences susceptible to development plans												Other influences										
Agriculture	Diverse agriculture	38	?	*	*	+	?				+	?	–		*		+	?	+	*	+	+	+		
	Land quality	39	?	–	–	–	?	–	–	*			–		?	+			+						
	Rural employment provision	40	?		+	+	–	+		–	+		–		*	–	+	?		*	+	+	*		
	Rural employment generator	41	?	*	+		?			*	+		–		*		+	?		*	+	+	+		
	Countryside management	42	?	*	–	+		*		+		?	–		–	+	+	?		*	+	+	–		
	Agricultural products	43	*	–	–	–		–		*	*		–		+	*	–	?	+	*	+	*	+		
Culture and recreation	Facilities for the arts	44		+	+		–				+						+	?		*		+			
	Heritage facilities	45		–	–	*	*	–	–	+	+	–	–		–	+	+	?	+	*		+			
	Sports facilities	46		*	+	+	*	+	–	*	+		–			+	+	?	+	*		+	+		
	Recreational facilities	47		*	*	+	*	+	–	*	+		–		–	+	+	?	+	*	*	+	+		
	Access and rights of way	48		–	*	–	*			+	?				–	+	+	?		*		+	+		

Valuable flora and fauna and habitats	Ancient woodlands	49	–	–	*	+	–	+	–	–	–	–	+	–	?	–	+	–	+	+	–
	Unimproved grasslands	50	–	–	–	+	–	+	–	–	–	–	+	–	?	–	+	–	+	+	–
	Wetlands	51	–	–	–	+	–	+	–	–	–	+	+	–	?	–	+	–	+	+	–
	Rivers/canals	52	–	–	*	+	*	+	–	–	–	–	+	–	?	–	+	–	+	+	–
	Open water/reservoirs	53	–	–	*	+	*	+	–	–	–	–	+	–	?	–	+	–	+	+	–
	Heathlands	54	–	–	–	+	–	+	–	–	–	–	+	–	?	–	+	–	+	+	–
	Mosses	55	–	–	–	+	–	+	+	–	–	–	+	–	?	–	+	–	+	+	–
	Geological features	56	–	–	–	+	+	+	–	–	–	+	+	–	?	–	+	–	+	?	–
	Upland moorland	57	–	–	–	+	–	+	–	–	–	+	+	–	?	–	+	–	+	+	–
	Areas devoid of wildlife	58	–	+				–				+	+	–	?					+	·
Archaeology	Historic towns	59	–	–	*	–	*	+	?	–	–	–	*	–	?	–	*	–	+	*	–
	Historic villages	60	–	–	*	–	*	+	?	–	–	–	*	–	?	–	*	–	+	*	–
	Uplands	61	–	–	*	–	*	+	?	–	–	–	*	–	?	–	*	–	+	*	–
	The marches	62	–	–	*	–	*	*	?	–	–	–	*	–	?	–	*	–	+	*	–
	River valleys	63	–	–	*	–	*	+	?	–	–	–	*	–	?	–	*	–	+	*	–
	Industrial archaeology	64	–	–	*	–	*	+	?	–	–	–	*	–	?	–	*	–	+	*	–

Table 11 The environmental influence of current policy areas in the West Midlands

INFLUENCES	CURRENT POLICIES	Housing — Supply of housing / A Maintenance of stock	B Supply of land	C Phasing	Location of new housing — D Concentration within existing settlements	E Peripheral expansion of existing settlements	F New settlements	G Concentration on public-transport corridors	H Reuse of land and buildings	The economy — Location of new / J Urban areas	K Rural areas
Climatic change	1	•	●		•	●	●	●	•	•	●
Built development pressures	2	•	•	•	•	●	●	●	●	•	•
Recreation and tourism development pressures	3		●		●	●	●	●			
Woodland planting	4		•			●	•	•		•	•
Costs and availability of energy	5	•	●	•	•	●	●	●	•	•	●
Mineral/aggregates (location and availability)	6	•	●	●	●	●	●	●	•	●	●
Land cost and availability	7	•	●			●	●	●	•	●	●
Green belt and other areas of restraint	8	•	●		•	●	•	●	•	•	●
Mobility/accessibility	9				•	●	●	●		•	●
Vehicle emissions	10				•	●	●	•		•	●
Pollution	11					●	●	●			●
Intensive agriculture	12										
Set aside	13										
Diversification	14		●			●	●	●	●		●
Financial (including levels of investment, grants and subsidies)	15	●	●	•	●	●	●	●	●	●	●
Water resources	16		●	•	●	●	●	●		●	●
Resource quality	17	●	●		•	•	●		●		
Agricultural land quality	18										
Public demand (including environmental education)	19	•	•						•		
Technical innovation (including recycling, etc.)	20	●						●	●		
Management of wildlife sites	21										

Table 11 (cont.)

	The environment								Minerals			Transport			Shopping		Waste disposal		Recreation/countryside access	
			Protect and enhance assets						Satisfy national/ regional demand			Accommodating traffic growth and reducing congestion			Location of new shopping facilities		Adequate provision for safe disposal			
	Green belt	Green wedges	Agricultural land	AONBs, AGLVs, SLAs, features	Wildlife habitats	Historical and archaeological features	New woodland planting	Greening	New materials	Recycled materials	Protecting known reserves	Major road building and road improvements	Traffic management	Encouraging greater use of public transport	Promotion and protection of existing town centres	Out-of-town retailing	Landfill	Promotion of recycling and alternative treatment of waste	Protection and proper management of existing sport and recreation facilities	Creation of new facilities
	M	N	O	P	Q	R	S	T	U	V	W	X	Y	Z						

Key:
● Increase
· Decrease

Table 12 The consistency of regional economic policies with the goals of regional environmental protection

Consistent policies	Inconsistent policies
• Regenerating the inner city • Improving the Housing Stock • Recycling land for industry • Environment improvements • Land reclamation • Tree planting • Community first initiatives • Maintaining a strong green belt • Protecting the best and most versatile agricultural land • Protecting the countryside resource • Concentrating higher-density housing and high-density employment in existing centres • Emphasis on maintaining existing centres • Nature-conservation enhancement	• Land-use policies that encourage out-migration from the conurbation and encourage commuting – particularly by car • The siting of business parks on green-field sites – on the periphery of the conurbation • Continued erosion of the countryside • Policies that sought to minimize the impact on the environmental resource rather than aiming to sustain environmental assets • Emphasis on road building rather than public transport • Allowance to build retail parks and out-of-town shopping • The emphasis on market forces on solving all problems

It is clear that market forces alone will not achieve the balance required and that a planned pattern of development and open spaces with an emphasis on sustainability (as now advocated in PPG12) is essential. Simply using the words 'sustainable development' is not sufficient to achieve sustainability. It is fundamental that future policy should actually prevent further air and water pollution and preferably lead to an improvement in these essential elements for life, and that the natural and man-made features of the rural and urban landscapes are not degraded further. This implies giving the opportunity and encouragement to the region's inhabitants to change to more sustainable lifestyles.

Objectives of an environment-led strategy

The primary objective of the environment-led regional strategy is to ensure that the region moves towards a sustainable environment in order to improve the quality of life. At the same time the regional strategy had to address the objectives of revitalizing the regional economy, providing for the region's housing needs, regenerating the main urban areas and tackling the social and economic imbalances which exist in the region.

However, the strategy must ensure that in meeting these wider economic and social objectives the intention of moving towards a more sustainable environment is met. Despite the proposals for an environment-led strategy, it was never the intention to prioritize any of the four objectives over the others. All must be achieved if the region has any chance of moving towards a more sustainable environment. Nonetheless, in order to achieve the primary objective it is necessary to pursue policies that will:

- improve the quality of air, water and land, and conserve finite resources;
- continue to protect and where possible extend amenity areas including the green belt, green wedges and other open spaces;
- protect and where possible enhance the region's tangible assets such as attractive landscapes, best agricultural land, historical and wildlife features;
- ensure that the region's consumable but finite mineral and energy sources are utilized more efficiently or alternatives found;
- satisfy housing and economic requirements while taking into account the environmental capacity of the region.

Regional planning guidance and environmental sustainability

The environment-led strategy was not the strategy submitted to the Secretary of State as the advice of the forum. However, elements of the strategy were incorporated into the advice. An extract from the advice report highlights this position:

> To work towards environmental sustainability as a vital ingredient underpinning the whole of this advice represents the major new force in regional strategic planning for the West Midlands. Sustainability is concerned not only with protecting existing environmental assets, although this is very important, but also creating new ones and with improving the quality of less favoured areas.
> In essence this means the:
>
> - notion that in satisfying our current needs, we should not compromise the ability of future generations to satisfy their needs;
> - idea that we should pass on to the next generation a package of assets of the same or greater value than that inherited from the previous generations;
> - premise that sustainability should not be achieved by unsustainable practices having to take place elsewhere.

The region's environmental assets do not exist in a vacuum separate from other planning considerations or global imperatives. Future planning in the West Midlands needs to fully recognize this in moving towards a more sustainable region.

Thus it is fundamental that strategic policies in the region should:

- aim to prevent further air, water and soil pollution and help improve their quality;
- seek to halt the denuding of our natural and man-made landscapes, rural and urban;

- recognize that economic growth and development need not be incompatible with moving towards a more-sustainable environment;
- realize that land-use transportation structures can influence the region's ability to combat pollution and conserve and enhance its environmental assets;
- make best use of existing infrastructure and spare capacity.

Limits to the environmental assessment process

Although the environment assessment process described above offers a real and practical way of enabling sustainable development by providing an environmental framework for economic and social regeneration, it does have a number of limitations.

In order to assess the experience of the West Midlands it is intended here to be highly critical but not necessarily comprehensive in terms of the shortcomings. The aim is to offer a critique that may be of benefit to others outside the West Midlands and not to concentrate on parochial concerns that have no wider implication. It is worth stressing at this point that the criticism should not be seen as a criticism of either the forum or the constituent authorities; it is understood that there are limits to the approach and more broadly as a consequence of current national structures and policies relating to the delivery of sustainable development. The critique covers six main areas, namely the political, financial and technical limits, issues of transparency and public involvement and matters relating to the continuation of the approach. The final part of this section will return to the key themes of the sources of pressure, the capacity to respond, the nature of the response, the performance of adopted measures and the scope for future progress.

Political limits

It would be wrong to imply that the forum accepted the conclusions of the SEWG report in its entirety. There were in fact some notable differences between the environment-led strategy and the advice prepared by the forum. Little purpose would be served by examining every difference of emphasis. However, there were two significant differences of view which could benefit from closer examination: the completion of the West Midlands Motorway Box, and the provision of land on the periphery of the conurbation for new industrial development.

SEWG proposed a range of public-transport improvements, together with planning policies that would minimize the need to travel. In addition, it concluded that new road building represented one of the largest threats to the region's environmental assets. However, SEWG did recognize that advice should '... If necessary make provision for limited new road building where there is no alternative for providing adequate access to remote rural areas or to improve the investment attraction of congested urban areas'.

However, within the context of a national transport-policy vacuum it would have been naive in the extreme to expect that the curtailment of the national road-building programme should start in your particular patch. It is currently a fact of life that a large number of investment decisions are still made by reference to road-based accessibility criteria. The decision by the forum to advocate the completion of the West Midlands Motorway Box was not taken lightly and it should be noted that it formed part of a package of other highly sustainable transport policies.

The second major difference from the SEWG proposals concerned premium industrial sites which were first proposed in PPG10. Not surprisingly perhaps, SEWG proposed that investment should be focused on brownfield sites within the existing built-up area close to where people live. This approach assumed adequate government resources for new infrastructure and site reclamation. It also recognized that it would take a significant amount of time to bring brownfield sites back into productive use. Given that central government had already proposed greenfield employment-land release in PPG10 and that the West Midlands over recent years had lost out to other regions who had provided new greenfield sites for inward investment (e.g. from Toyota, Siemens and Nissan) it is perhaps not surprising that the forum adopted a different stance from that advocated by the SEWG.

A level playing field?

It is a basic principle of sustainable development that it should not be achieved at the expense of neighbouring areas. The emergence of a global economy where individual regions around the world are competing for new economic investment will inevitably lead to more imaginative definitions of what is meant by sustainable development. To attract new investment, regions world-wide will move heaven and earth and every known sustainable principle to attract new investment from outside their region. Competitor regions are prepared to offer almost any site to attract such beneficial fish and unless the West Midlands follows this lead it will fail, losing out on new job opportunities for the people of the West Midlands.

Clearly, the SEWG proposal to reuse brownfield sites and create new high-profile and attractive investment sites is preferable. However, it is recognized that this takes time and resources. If a level playing field existed, together with adequate resources, the short-term benefits of greenfield-land release would be totally unacceptable. However, that sustainable planning system does not yet exist.

The speed of change

As mentioned above, the forum included within its advice a package of employment and transport/accessibility policies that are sound and sustainable proposals. The fact that there are some proposals that favour short-term benefits for this generation which will probably disadvantage future generations should not disguise the fact that overall the advice will move the region towards a more-sustainable environment. The direction of change has never been questioned by the forum, what has been debated is the speed at which change takes place.

Some, no doubt, would argue that the forum proposals do not go far enough to force the pace of change. However, it is important to note that it is democratically elected representatives (whether indirectly or directly elected) who choose the pace of change. Sustainability cannot be achieved by a sudden change in direction which unreasonably sacrifices the economic needs of this generation. It is preferable to move at an acceptable pace that begins to balance the needs of the present with the needs of the future and to maintain democratic control over the process.

It is possible to conclude that the forum's advice to the Secretary of State amounted to nothing more than a coalition of parochial interests masquerading under the banner of regional consensus. Such a conclusion would be wrong, however. The forum did make a number of key decisions that limited short-term economic gain by taking into account longer-term concerns for sustainable development. A good example of this approach was the general proposal to shift the emphasis of development from the south-east of the region to the north-west. While it is true that such proposals meet existing sub-regional aspirations, this shift in emphasis also recognized the need to maximize the development potential of the West Midlands conurbation, which in the ultimate analysis means a shift in public-sector resources as well. It is undoubtedly true that parochial concerns were subsumed for the benefit of long-term sustainable development and the region as a whole.

Financial limits

SEWG had little or no responsibility for allocating resources to implement its proposals. The forum on the other hand contains within its membership the leading local-authority decision-makers in the West Midlands who control collectively billions of pounds of public resources. However, both SEWG's proposals and the forum's advice assumed the redirection of resources largely beyond their control. This situation begs the question to what extent should strategies that seek to match economic development with environmental management be prepared in isolation from financial realities?

To some extent SEWG was too divorced from financial realities. However, in defence it can be argued that to limit any vision to the resources that exist at any one time is simply a recipe for inertia. It simply fosters a 'can't do' philosophy. In addition, although the SEWG's ignorance of current financial realities may have been profound, it is equally fair to argue that current accounting systems are ignorant of longer term and environmental costs, something that certainly was considered.

The forum did, however, attempt to cost the implementation of its advice, but could not help but conclude that not only should resources be redirected but additional public-sector investment was necessary. It is true to some extent that the forum's advice did act as a resource-bidding document which is hardly surprising given the current perceived underfunding of local government. Whilst it may be unpalatable for some, it is difficult to conclude that sustainable development can be achieved at the same initial cost as short-term non-sustainable proposals. Such conclusions raise a fundamental concern in that will the lack of resources simply slow down the pace of change or can it actually undermine the totality of the strategy?

In order to examine this dilemma two examples can be cited. First, both SEWG and the forum agreed that at the heart of any strategy that seeks to move the region towards a more-sustainable environment there should be a commitment to urban regeneration. Second, there was the dilemma over how to accommodate the conurbation's 'overspill' population. The first example demonstrated that for urban regeneration to be successful additional resources will be required. Without additional resources, decentralization trends will continue eroding environmental assets outside the conurbation and denuding those within. To some extent a strategy of containment will succeed in any event, simply by controlling the supply of housing beyond the conurbation. This ghetto mentality is certainly not what was envisaged by the forum. It may sound like a cliché but it was both SEWG and the forum's strategy to create a conurbation where people want to live not where they have to live. Unless public resources are made available for urban regeneration projects (beyond that currently available to forum members), it is difficult to imagine how the massive task of creating a vibrant and quality urban environment can be achieved. Containment may well be realized, but the goal of a sustainable quality environment for nearly half the region's population will be left as simply a pipe dream.

Despite SEWG and the forum's objective to produce a conurbation where people want to live there will always be a degree of decentralization; the so-called urban 'overspill'. Reorganizing this reality SEWG considered two options. First, they considered the encouragement of balanced communities beyond commuting distance of the conurbation where jobs and homes are generally matched. Second, it looked at the creation of growth corridors radiating out from the centre of the conurbation based on public transport (primarily rail) corridors. It was SEWG's conclusion that since the forum was not in a position to control investment in public transport, it could not risk proposing the growth-corridor option. Lack of investment in public transport could simply lead to elongated car-based commuting corridors. SEWG therefore chose the balanced-communities option which was felt to be less risky. The forum, however, placed more emphasis on growth corridors than on balanced communities.

While it is hoped that investment will take place in public-transport provision within the corridors, it certainly cannot be guaranteed. While this option may be the best available, it does highlight the problem of developing strategies that are divorced from financial responsibilities and the consequent risks that are involved.

Technical limits

The membership of SEWG contained experts from a wide range of organizations, but interestingly not from NGOs, which in retrospect was a mistake. However, the inclusion of experts from outside the local authority was in our view a great asset. It is no understatement to say that the task could not have been completed without them. That is not to suggest that SEWG's knowledge was perfect, it was far from it! At the time of compiling the SEWG report (1992) there was little relevant published material available and SEWG was thus required to develop its own methodology and information sources.

Limitations in this respect were many, most notably regarding air pollution and renewable energy resources. SEWG became increasingly frustrated by the lack of meaningful information in these areas, but rather than delay the whole process it made the best of what it had and then admit that it did not have all of the answers.

Public involvement

Although the work of SEWG was open to public scrutiny and the Regional Planning Guidance Conference was held in public, public involvement was limited. It has been a growing practice of the forum to develop partnerships with a range of organizations that represent a particular view or interest. These groups range from the Church of England to the House Builders Federation. Increasingly, real and meaningful partnerships are being formed and it would be wrong to criticize the lack of public participation in preparing the forum's advice without giving due credit to a genuine desire by the forum to involve others in the decision-making process.

However, it remains a basic fact that the public of the West Midlands were not involved in any meaningful sense with the planning of their region. Of course, compared with other planning guidances issued by the Secretary of State, there was greater public involvement, but compared with structure plans or UDPs little was achieved. Indeed it is not clear that it is a practical suggestion that public participation should be attempted at the regional level. What is clear is that the more site-specific regional planning becomes (which could be a reality if regional government ever occurs), a greater effort will be required to engage the public in the decision-making process. It is ironic that the Local Agenda 21 process has led to a quiet revolution in public involvement in local environmental decision-making, but has not been attempted above the county level. If it were to be attempted, new tools will be required and a recognition of additional resources and time must be accepted.

Transparency

One of the successes of SEWG's work was the transparency of the reasoning behind its conclusions. The three matrices exposed the outcome of every element of the discussions which enabled anyone to challenge the findings. The report of SEWG's work was equally open allowing for a step-by-step understanding of the process. By adopting such an open attitude there is a clear requirement to argue the case for a particular approach and not simply assume that there is only one sustainable solution. Equally it is vital that when information is not available that this is freely admitted and not obscured.

Continuing the process

At this point in time it is not clear if the West Midlands is moving towards or away from a sustainable environment. The work of SEWG should have been immediately followed through with the development of an environmental monitoring system for the region. To some extent the momentum has been lost and it is only now that a regional environmental monitoring group is being established over a year after RPG for the West Midlands was published.

It is important not to give an impression that progress has not been made. Due to the novelty of regional environmental monitoring and regional environmental reporting, the forum decided to examine good practice in the rest of the European Union (EU). In order to achieve funding from the European Commission (EC), it was further agreed to promote an EU project to be co-ordinated by the Association of European Industrial Regions (RETI), with partners from non-UK countries. The final project included four partners; that is, Asturias (Spain), Greater Manchester, Verona (Italy) and the West Midlands. The West Midlands (specifically Shropshire County Council) agreed to lead the project and two consultancies were commissioned in order to look at good practice.

The objectives of the Regena Project were to:

- develop an integrated approach to environmental information from measurement and gathering of data through to the presentation of information;
- development and use of standardized measuring techniques, harmonized definitions and harmonized reporting;
- applying emerging advice on environmental indicators and developing a meaningful collaboration with the European Environment Agency.

Clearly, the Regena Project covers objectives beyond the original requirements of the forum. However, the key requirement of providing good practice on environmental reporting has been achieved and included within the final report.

Regional planning guidance for the West Midlands

It is not the purpose of this chapter to examine the contents of the final regional planning guidance issued by the Secretary of State for the Environment. However, it is gratifying to record that with some notable exceptions the

Secretary of State did accept the forum's advice. Sustainability remains a central theme of the government's vision for the West Midlands.

The tough question is what was excluded. In general, any proposals in the environment-led strategy that involved new and additional resources were excluded. Not surprisingly perhaps, proposals that involved the radical change of existing policy were also excluded, including the encouragement for the fuller occupancy of under-used and vacant housing.

Regarding the economy, the forum concluded that no further peripheral high-quality sites should be identified in regional planning guidance over and above the PPG10 sites and that effort should be focused on the regeneration of brownfield sites. This conclusion was not accepted either by the Secretary of State for the Environment and new greenfield employment sites were proposed in the form of major investment sites.

Although two areas have now been identified by the forum for major investment sites one is now subject to a consideration by a planning inspector. The inspector's report is eagerly awaited! The conclusion by SEWG that the completion of the West Midlands Motorway Box was not desirable, was also not accepted. It is ironic that the government has now decided not to pursue this objective, despite its inclusion within regional planning guidance. One final and yet-to-be-explained difference was the exclusion of all references to environmental capacities (which were included in the draft RPG).

Conclusions

The development of an improvement in the integration of economic development and environmental management in regional planning in the West Midlands emerged in line with the profound developments taking place in the outside world during the beginning of the decade and before. Without doubt the biggest influence at the time of preparing the SEWG report was the Brundtland Report (WCED 1987).

At the end of 1990 the UK Government published the white paper, *This Common Inheritance: Britain's Environmental Strategy*. This white paper made it clear that not only should change take place but that it could happen and that the town planning system had its part to play in achieving sustainable development. The subsequent agreements achieved at the United Nations Conference on Environment and Development – the Rio Earth Summit (1992) and the development of sustainable development – the UK Strategy (1994) endorsed the SEWG's emphasis on sustainable development. Notwithstanding the UK Government's changing view on economic development and environmental policy, the development of thinking in the European Union, influenced not only the broad approach but also the detailed proposals. For example, the drafts of the EU's Fifth Environmental Action Programme, entitled *Towards Sustainability*, reached us before the Rio Earth Summit and helped to frame its understanding immensely.

The work of SEWG helped to shift the debate in the West Midlands away from considerations of unsustainable growth to a position where discussion took place within the parameters of sustainable development. How these parameters are defined will be continually debated. The most fundamental point to note is that while the extent of economic and environmental integration were strongly argued, the need to consider development options had to be and were justified within a sustainable set of arguments. Whatever the justifications of their arguments, the House Builders Federation considered it advantageous to argue for new build on the grounds of energy efficiency and global warming!

In the West Midlands, as in all other English regions, competence in spatial planning rests with the local authority and central government. The West Midlands local authorities have a long history of co-operation. There has also been a longer history of local difficulties. The opportunity for local authorities in the West Midlands to influence the content of government guidance for the West Midlands was recognized very early on. The West Midlands Forum of Local Authorities possesses only a very small bureaucracy, but has a very strong and articulate political body. This situation, perhaps inevitably, led to working arrangements whereby leading local authorities were identified to carry out the work of preparing advice for the consideration of the forum's politicians. The twinning of shire county with metropolitan authority not only retained a political balance, but allowed for different perspectives to be heard.

The recognition of the need for a greater environmental input into the forum's decision-making process has resulted in long-term changes. Since the publication of SEWG's report the forum has responded to the imperative of moving towards a more sustainable approach in a number of distinct ways. A new Environmental Members Panel has been established, a Sustainable Regional Transport Strategy is being prepared and the process of identifying how the conurbation's overspill population is to be housed has moved a long way to considering environmental capacities (even if such considerations are excluded from regional planning guidance) and the long-term sustainable implications of locational decision-making.

It is still too early to draw any real conclusions on the performance of the measures adopted in regional planning guidance. The concerns expressed by SEWG concerning matters beyond national planning policy, at the time, are now being addressed. The lack of integration between transport and land-use planning at the regional level, for example, are specifically being addressed in a revision to PPG13 and indeed through the new central-government structures introduced by the Labour Government.

The West Midlands Region is unique; its central core in the West Midlands conurbation allows for the development of novel sustainable land-use strategies. While the political structures, and spatial characteristics of different regions must determine specific policy approaches, it is now felt that the SEWG approach, with its matrices, its joint approaches to working and its focus on environmental assessment is relevant beyond the West Midlands.

SEWG provided a framework for the forum in which economic and environmental objectives could be considered in a truly sustainable way. Much of

SEWG's proposals were accepted by the forum and it could be concluded that the forum did look at both the positive and the negative outcomes of the environmental assessment when drawing up its advice to the Secretary of State.

Endnote and acknowledgement

The Sustainable Environment Working Group was established by the West Midlands Forum of Local Authorities originally under the co-chairmanship of Chris Carter of Hereford and Worcester County Council and Paul Watson of Dudley Metropolitan Borough Council. Steve Winterflood, also of Dudley MBC, subsequently replaced Paul Watson. The views expressed in this chapter are the personal views of Chris Carter and Steve Winterflood and not necessarily those of their employing authorities or the forum. Thanks are due to Rona without whom none of this would have been possible.

Reference

WCED (World Commission on Environment and Development) (1987) *Our Common Future*, Oxford: Oxford University Press.

10 Strategic frameworks for local environmental-policy integration

The case of Leicestershire

Alexander Johnston

Introduction

Within Leicestershire the broad strategic context for economic development and environmental action is set by the Leicestershire Structure Plan. Not only are structure plans statutory documents, with all the status associated with that designation, but they also provide one of the few real opportunities for elected members to address long-term strategic issues. The fact that we have a 'plan-led' town and country planning system (under the provisions of Section 54A of the Town and Country Planning Act 1990), with the first point of reference for all promoters of development being the policies and proposals contained in a development plan, adds to their central importance.

Leicestershire County Council members led the preparation of the current structure plan. They achieved this through a series of member working-party meetings. Indeed, certain members actually participated in the examination in public of the plan. Leicestershire County Council members are, therefore, more than comfortable in dealing with strategic planning issues. They have ownership of the plan. They appreciate its central role in guiding the preparation of the many other policy and programming documents prepared by the county council, district councils and others.

This capability and awareness has undoubtedly been assisted by having a combined planning, economic development and transportation department. Responsibilities cover strategic planning, highways, important aspects of public transport, economic development and training, environmental improvement, conservation and waste regulation and disposal. Transport engineers, public transport co-ordinators, economic developers, waste regulators and environmental improvers/conservers, have all had the opportunity to contribute to the development and implementation of strategic planning policy.

Although the major responsibilities lie within the Planning and Transportation Department, many areas of work − such as the preparation of the county council's Rural Strategy and Action Programme − have important corporate and inter-agency dimensions. The lead on such corporate matters is invariably taken by the Planning and Transportation Department. The process has benefited from good working relations with the district councils, the Government Office for the East Midlands (GOEM) and other agencies such as the Rural Development Commission and the National Forest Development Company. District councils and GOEM attend regular meetings with the county council on strategic planning issues where they have the opportunity to contribute to the formulation of strategic planning policy. Regular liaison takes place with developers under the aegis of the Builders' and Developers' Forum.

This chapter describes the Leicestershire Structure Plan and other strategic policies in outline before considering how it is being implemented. It then provides a brief assessment of the Leicestershire experience in moving towards a more-sustainable future with particular regard to the interface between economic growth and environmental conservation.

The structure plan and other strategic policies

This section describes the Leicestershire Structure Plan, the county's transport strategy and other main areas of strategic policies.

The structure plan

The current Leicestershire Structure Plan was adopted by the county council in January 1994. Although there are many similarities with the previous structure plan, the new plan places greater emphasis on transport choice, sustainability and a more selective approach to economic development. It is consistent with the regional strategy prepared during the same period by the East Midlands Planning Forum. This consistency was assisted by Leicestershire County Council being the 'lead' authority during preparation of the regional strategy, which the Secretary of State has had regard to in preparing regional planning guidance for the East Midlands, subsequently published as Regional Planning Guidance Note 8 in March 1994. The structure plan consists of a written statement of the county council's policies and general proposals of structural importance for the development and other use of land in the county. It also provides the framework for more-detailed development-control policies and proposals in local plans.

In the structure plan the county council has sought to create a strategic planning framework designed to accommodate economic growth, to improve employment opportunities and increase economic prosperity, whilst giving increased emphasis to a better quality of life through the conservation and enhancement of the environment. The central component of the structure plan is the county council's innovative Transport Choice Strategy, which underpins the overall strategic framework of the structure plan including the transport policies.

Transport choice

In June 1989, the county council began work on 'transport choice', a fundamental review of transport policy for Central Leicestershire, to the year

2006 and beyond. Initial appraisal of a number of options confirmed that Leicestershire's continuing prosperity will rely on the availability of a transport system able to meet the increasing demands placed upon it. The extent to which this can be achieved by increasing car use is limited for environmental reasons and the physical capacity of the road network. As public transport is more energy efficient and environmentally friendly, it could be exploited as a more acceptable alternative to ever-increasing car use. As a result, the county council adopted its Transport Choice Strategy in which the central objective is to create more capacity in the transport system by providing the choice of high-quality public transport as an alternative to car use. This work was initiated and developed before the publication, by the Department of the Environment (DoE), of PPG13 (Transport) and PPG22 (Renewable Energy).

The Transport Choice Strategy does not seek to force a reduction in car use by punitive means. It seeks to improve options for people making journeys in both the short and long term so that they are not forced to use cars because there is no other option available to them. The intention is also to reduce the pressure of private motor-vehicle movements on the environment and improve the quality of life of Leicestershire residents. The strategy aims to improve bus, train, cycling and walking options by the allocation of land where such possibilities already exist or may be sustained or provided by the concentration of development.

The Strategy also involves action by the county council and others to support the improvement of the local rail network, measures to encourage better quality and more efficient bus services and cycling and walking. In this way a more energy-efficient transport system will be achieved.

Location of development

The structure plan, therefore, proposes a settlement pattern which supports public transport in order to assist the provision of an effective alternative to the private vehicle. Most people in the county live in urban areas and, for them, improved bus services offer the most effective way to provide transport choice, thus reducing reliance on the private car for many types of journey. Between urban areas new railway services proposed by the county council will provide the frequency and quality of services to provide transport choice. Bus services will only provide this degree of choice for inter-urban journeys where services are very frequent, and where the densities of built-up areas enhance profitability.

It is proposed that most development will occur within and adjoining the urban areas of Leicester and adjoining settlements of Ashby, Coalville, Hinckley, Loughborough, Lutterworth, Market Harborough, Melton Mowbray, Oakham, Shepshed and Uppingham. Of the smaller amount of new development which will occur elsewhere, most should be in locations along the transport corridors which offer or will offer a realistic choice of transport, once proposed new services are introduced. The corridors are based on the railway lines in the county, along which new local railway services are proposed, and along the A6 between Leicester and Rothley and between Quorn and Loughborough, including sections of the former A6 between Rothley and Quorn. This is the only bus route in the county which has the potential to offer, in the foreseeable future, a realistic choice of transport.

Development will take place at nodes along the corridors. It is not proposed that ribbon development will take place along these corridors. Other policies of the plan protect the countryside, green wedges and the separation between settlements. Development land will be allocated in locations within walking distance of a station or proposed station or the A6 bus route (walking distance is usually taken to be about 1 km). There may be exceptional circumstances where the opportunities for park and ride allow development land to be allocated further away.

Rail services

The Transport Choice Strategy, therefore, seeks to exploit the potential of rail in providing a viable alternative to use of the car by establishing rail services on all five rail routes linking Leicester with the county's main towns. With enhanced train services on all major routes in the county, and with complementary land-use policies seeking to locate new development close to stations, rail can play a major part in delivering transport choice. It is the county council's intention to invest to secure the introduction of a network of train services. This will include the construction of twenty-four new stations and associated car parks.

Rail transport is to be encouraged to move freight. Structure-plan policies seek to ensure that developers of major new employment areas utilize the potential of rail. Rail transport will be sought particularly in the movement of minerals, as set out in the Leicestershire Minerals Local Plan. The views expressed in the East Midlands Regional Strategy that the provision of a number of rail freight terminals in the region are required to serve the whole of the East Midlands is supported by the county council and indeed proposals have now come forward from the private sector.

Buses

Given the right operating conditions, buses can offer a cost effective and flexible means of transporting large numbers of people efficiently, particularly within urban areas. The structure plan proposes that bus services be given the opportunity to establish themselves as a viable alternative to the car. This will be achieved if new development is properly designed and laid out to allow buses to penetrate and serve new areas effectively. The possibility of providing additional infrastructure offering buses further competitive advantages, such as bus lanes, bus-only gates and bus priority at traffic signals, will be explored. The funding of infrastructure will normally be borne by developers within national guidelines, currently provided by DoE Circular 16/91. Developers will not be expected to subsidize bus services over the long term because one of the main principles of transport choice is that services should be viable and self-supporting.

In certain circumstances, however, it may be necessary for developers to offer subsidies over a limited period of time.

These measures, coupled with a continued investment programme by the county council in bus infrastructure and joint action with bus operators to improve service quality, are intended to secure buses a much larger and growing share of movements. Close and direct access for passengers to the buses they wish to use, at both ends of their journey, is vital for the success of the strategy.

Other strategic considerations

Strategy Policy 1 of the structure plan sets out other key considerations. They include measures to:

1   conserve and enhance the environment, minimize built development in the countryside, protect the best and most versatile agricultural land, and assist the development of the National Forest;
2   provide a continuous supply of housing land, release a range of sites for employment development including high-quality employment sites for B1 and B2 uses and take measures to ensure a mix of housing and employment uses in each locality;
3   regenerate the priority areas (Leicester's inner area and the former coalfield areas of North-West Leicestershire), recycle derelict land and seek better use of vacant and under-used land and buildings;
4   enhance the role of city and town centres particularly with regard to their shopping role;
5   reflecting Leicestershire's role as a major minerals-producing county, to provide a strategic context within which minerals and waste-disposal proposals can be considered.

These measures will now be examined in turn.

Conservation and enhancement of the environment

The thrust of these policies is to improve the urban areas, strategically influence urban form, protect the countryside, encourage reclamation and effect the implementation of the National Forest. Leicestershire County Council played a leading role in attracting the National Forest to the Midlands. The National Forest was initially put forward by the Countryside Commission in 1987 in its policy document, 'Forestry in the Countryside'. The Countryside Commission announced in October 1990 that it was recommending to the Secretary of State for the Environment that the National Forest should be established in the Charnwood–Needwood area. The Secretary of State subsequently announced that the project had his support and could proceed.

The forest will comprise about 50,000 ha with a mix of wooded areas, open country and farmland. It will be a working forest and a vehicle for large-scale

environmental improvement. Opportunities will be provided for inward investment and eventually the forest will contribute to the production of home-grown
timber. As such, it has been appropriately described by a former Chairman of the
Countryside Commission, Lord Barber, as '... the most challenging and
potentially rewarding environmental project of the 1990s'.

Housing and employment

As well as encouraging the release of land for residential development within or
on the edge of the main urban areas, at settlements adjoining Leicester and
locations along the transport-choice corridors within walking distance of a station,
or proposed station, or the A6 bus route, the plan encourages the release of small
sites for housing within urban areas, and the economical use of land through
higher-density development, given that single-person and single-parent households, as a proportion of overall needs, are expected to increase significantly.

Similarly, most of the 1035 ha of employment land to be provided between
1991–2006 will be allocated in locations which offer, or will offer a realistic choice
of transport. The only major exception to this policy is the release of land to meet,
to a limited extent, the demand for economic development around junctions
23a and 24 of the M1 (M42–East Midlands Airport junctions). The type of
development proposed is a high-quality employment site (HQES) and a major
distribution facility. Three additional HQESs and a further major distribution
facility are also proposed but all will be sited in locations offering a choice of
transport mode. HQSEs are intended to provide accommodation to retain and
attract companies with long-term growth potential to the county. Occupiers of
the HQSEs will be restricted to B1 and B2 uses (Use Classes Order 1987).

Policies and proposals are also contained in the plan to enable provision of a
science park in Leicester, the further diversification of the Leicestershire economy,
protection of employment land and buildings so that they are not lost to other
uses, and selective and sensitive development in villages in order to bolster the
rural economy.

Priority areas and recycling derelict land

A common theme running through the plan and reflected in the wording of many
of the policies is the priority and importance to be given to the regeneration of the
priority areas of the inner area of Leicester and the area of mining decline centred
on North-West Leicestershire. Most of Leicestershire's derelict land is located in
the former deep-mining area of North-West Leicestershire. That is where the
county council's reclamation will be concentrated but not to the total exclusion of
reclamation activity elsewhere.

City and town centres

The emphasis of the policies on city and town centres is to sustain and increase their shopping role and their importance as professional and financial centres. Such a strategy of concentration is more sustainable than one of decentralization. Provision is to be made in central areas for residential development combined with improvements to the safety and attractiveness of the environment, particularly for pedestrians and cyclists. The previous structure plan contained a policy which sought to sustain and increase the sub-regional role of Leicester City Centre in respect of durable goods' shopping. This policy was a central consideration in the eventual refusal of planning permission for Centre 21 (a regional shopping centre of 1.2 million sq ft proposed to be located on the periphery of Leicester at the junction of the M1 and M69 motorways), by the Secretary of State in July 1988 (Sabey 1988; Johnston 1988).

Subsequently, a major resurgence of new investment in shopping, office development and city-centre environmental improvements, has taken place once the 'shadow' of Centre 21 was lifted from the city centre. This investment has been co-ordinated and initiated by the Leicester City Centre Action Programme, prepared and implemented jointly by the city and county council. The existing structure plan extends this protective and enhancing planning framework to the county towns of Leicestershire, although, in certain circumstances, new out-of-town retailing and, to a more limited extent, office development, may be acceptable.

Minerals and waste

Leicestershire is one of the largest and most diverse producers of minerals in the UK. Its wide range of minerals includes sand and gravel, granite, various types of limestone, gypsum, ironstone, various types of clay, opencast and deep-mined coal and oil, some 1,630 jobs directly depend on the industry. The winning and working of minerals is environmentally damaging. New proposals are often highly controversial.

The principal objective of Leicestershire County Council's Minerals Planning Authority is, therefore, to balance its responsibilities to ensure a supply of minerals for the needs of the community with its responsibilities to keep the environmental impact and other effects of mineral extraction to an acceptable level. The policy framework to ensure the maintenance of such a delicate balance is provided by the broad-brush strategic policies of the structure plan supplemented by the more detailed policies and proposals (including the identification of specific sites) contained in minerals local plans.

Up to April 1996, the county council had three important responsibilities relating to waste: it was the Waste Planning Authority, the Waste Regulation Authority and the Waste Disposal Authority. The second of these responsibilities – waste regulation – was subsumed within the new Environment Agency, along with the former National Rivers Authority and the Inspectorate of Pollution, in April 1996.

Before the transfer of responsibility, the county council, as Waste Regulation Authority, prepared its Waste Disposal Management Plan (this plan is discussed in more detail later) which, along with the structure plan, sets the strategic context for the treatment of waste arising in Leicestershire. The translation of this broad strategy into proposals involving the use of land (for composting plants, waste recovery, waste transfer, incinerators and landfill) is the responsibility of waste local plans.

Waste planning, particularly the identification of landfill sites is, perhaps, even more controversial than proposals to win and work minerals. When proposals to extract minerals incorporate proposals to fill the resultant voids by infilling of waste – as happened recently in the case of the New Albion proposal in the Ashby Wolds area of Leicestershire – then the issue can become highly charged and politicized and the decision-making process somewhat extended.

Strategic policy implementation

The mechanisms

Various mechanisms and instruments exist which allow for the implementation of policy. Chief amongst these are local plans, transport policy, environmental programmes, the areas of mining decline strategy, rural policy action, economic initiatives and waste-disposal plans.

Local plans

Figure 3 indicates the means by which strategic policy is implemented in Leicestershire. Although strategic policy is implemented by a variety of public and voluntary agencies and the private sector, local plans provide the main means through which the policies of the structure plan are implemented. Local plans are statutory documents and district councils have the main responsibility for their preparation. The nine district councils in Leicestershire are preparing new plans for their areas in the context of the structure plan. These plans are well-advanced: the City of Leicester Local Plan has been adopted by the city council and full local-plan coverage was in place in 1997.

Once this is achieved, Leicestershire will have comprehensive local-plan coverage for all its area for the first time. In addition, the county council has prepared and adopted its Minerals Local Plan (May 1995) and is preparing its Waste Local Plan. Adoption is expected in 1997.

Transportation

The county council bids formally for government support to implement its transport policies by submitting annually a Transport Policies and Programme (TPP). Although the TPP is not a mandatory document, it is required to be consistent with the structure plan. As part of the 1994/5 bid, the county council developed a single 'package' programme covering both road and public transport investment proposals in the Central Leicestershire urban area. The 1995/6 bid

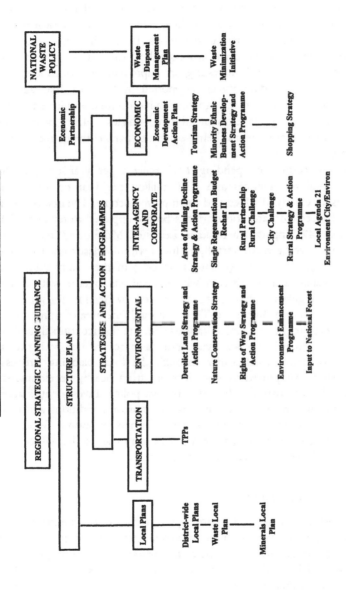

Figure 3 Strategic policy and implementation structures in Leicestershire

developed this process further. Proposals need to demonstrate how they link into specific land-use policies and to show that full consideration has been given to the potential for public transport, cycling and improved traffic management as alternatives to new road construction, as anticipated by the structure plan.

Action to secure the provision of bus services to new development is ongoing through the inclusion of policies in district local plans and negotiations with developers regarding developer contributions. Partnerships are being developed with the bus companies in order to improve the quality of bus services. As part of this, the county council is investing in improvements to routes, junctions, information systems and terminals, supported by the TPP 'package funding'. Major bus companies are already showing an increasing commitment to their side of the bargain through investment in new vehicles, quality programmes and training. Further commitment is being sought to parallel the council's investment programme.

The county council is spending £1.2 million per year on providing and promoting local bus services and £3.0 million per year on providing travel concessions for elderly, disabled and unemployed people. The first of a number of planned enhanced bus-lane corridors has been completed along one of the main radials – the A47 Uppingham Road in Leicester. Monitoring indicates that bus journey times have been successfully reduced without detriment to other traffic.

The first stage (Leicester to Loughborough) of the Ivanhoe Line has been opened. This stage includes three newly built stations together with an additional platform at Loughborough station, all assisted by Section 56 grant. Stage 2, between Leicester and Burton-on-Trent (through the area of mining decline), is currently being progressed. Consultations with the district councils and others to identify and safeguard sites for these stations in accordance with the structure plan, are proceeding satisfactorily through the local planning process. Their provision will be supported by contributions from nearby developments where appropriate, in accordance with structure plan policy. Investment is accelerating in traffic-calming schemes in urban areas to improve conditions and routing for pedestrians and cyclists, and in the provision of new cycleways. Good progress is also being made in developing a county-wide lorry plan for heavy-goods routing, again in accordance with structure plan policy.

The environmental programme

'Leicestershire Landmark' is the name given to the county council's comprehensive programme of environmental improvements. The programme is strategy driven: emphasis is given to partnership working and leverage. The county council's financial contribution of about £0.6 million is augmented by monies from a variety of other sources thus supporting a programme of some £2 million. Derelict-land reclamation is the programme's largest component. Over the last several years this programme has increased from some £50,000 to about £1 million per year. This work is guided by a derelict land strategy and action programme which is regularly updated and, in turn, is guided by the structure plan. The main current project is the reclamation of Donisthorpe Colliery and Tip as a woodland park within the National Forest. This major project has attracted funding from the Rural Development Commission, the European Union (RECHAR 11) and English Partnerships.

With regard to nature conservation, the structure plan policies are important considerations to be taken account of in the control of development and in allocating specific sites for development in local plans. As well as this regulatory approach, however, a positive programme of action has been approved through the Nature Conservation Strategy. A nature-conservation forum has been established and initiatives being progressed include the creation of roadside nature reserves, two countryside management projects focusing on the area of mining decline and the Leicester urban fringe, and measures to increase public awareness such as training courses for volunteers in rural skills and a programme of guided walks. Experience in Leicestershire suggests, perhaps surprisingly, that potential conflict between economic development and nature conservation can often be minimized by careful planning.

A similar approach characterizes the county council's initiatives on the rights-of-way network. In December 1995, the county council approved the 'Milestones Statement', which contains a summary of the work undertaken towards meeting the national target of a rights-of-way network which is legally defined, properly maintained and well publicized by the year 2000. Leicestershire is one of the few county councils expected to meet this target. The council consider the development and promotion of the network will foster 'green tourism' thus assisting diversification of the rural economy.

The Environment Enhancement Programme consists of three smaller pro-grammes: the Building Conservation Programme, the Landscape Programme, and the Environment Improvement Programme. All are guided by the priorities set by the structure plan. Building-conservation grants and specialist advice are offered to encourage owners of historic buildings to carry out repairs using appropriate techniques, and where appropriate, recycled materials. Enhanced grants are offered in 'town schemes' and Conservation Area Partnership towns, with help from English Heritage and district councils. During 1995/6 more than 400 historic gardens and landscapes have been reviewed with the aim of developing their tourism potential.

Also noteworthy is the improvement by the county council of the environment of Castle Park, the impressive historic core of Leicester. A significant new open space has been created; the historic Castle Gardens have been remodelled with the creation of a story-telling area, the erection of a maypole, the opening up of views of the River Soar, along with comprehensive signage and interpretation. This will further improve the attractiveness of the city centre for visitors and shoppers.

The Landscape Programme assists landowners and community groups to carry out projects involving tree planting, hedge planting and laying, renovating drystone walls, creating or renovating ponds, pollarding waterside willows and managing small areas of woodland. Of particular importance to the county is the Leicestershire Hedgerow Project. A project officer, partly funded by the

Countryside Commission, gives advice and assistance to landowners on the planting of new hedgerows and maintaining existing plantings. Attractive grants are available and compared with most other counties the length of hedgerow in Leicestershire is increasing impressively, albeit from a relatively low base.

Finally, the Environmental Improvement Programme is directed towards a wide range of projects. One of the largest and most interesting is the reinstatement of the Ashby Canal to the north of Measham to its historical terminus close to Albert Village at the Derbyshire border. Implementation of the National Forest continues apace. An indicative strategy and implementation plan has been prepared by the National Forest team. Since the National Forest was designated 5 years ago, about 1 million trees have been planted, the majority of which are in Leicestershire, on 130 sites ranging from small farmland planting schemes to major land-reclamation projects involving 50,000 or more trees. Excellent working relations are maintained with the National Forest team.

Area-of-mining-decline strategy and action programme

By the late 1980s the Leicestershire coalfield was nearing exhaustion. The mining industry employed more than 7,000 people in North-West Leicestershire in 1981: by 1987 only 3,000 people worked in the industry declining to about 1,000 by 1990. The local environment had been severely damaged by both coalmining and large-scale quarrying and clayworking. Very little industrial development had taken place and the widespread effects of subsidence meant that little good-quality land for industrial development was available. The area had a poor 'image'. These problems were exacerbated by significant and unexpected job losses in other industries, and by the mining rundown being quicker than British Coal originally forecast.

In 1981 the county council and the two district councils representing the area founded a tripartite committee to stimulate and co-ordinate measures to counteract the effects of mining rundown. A three-pronged strategy was adopted focusing on economic, social and environmental regeneration. It was anticipated that unemployment by 1989 would range from 22 to 29 per cent. Unfortunately the area received little in the way of central-government assistance. Requests for assisted-area status, and for the rural parts to be designated an RDA were rejected. The area was, however, declared a derelict-land clearance area, thus attracting 100 per cent grant on derelict-land schemes.

The councils' commitment to the regeneration of the area continues. The Tripartite Committee continues to oversee the implementation of the Regeneration Strategy and Action Programme, which is regularly updated. Massive investment, both public and private, has been directed and attracted to the area. The rate of unemployment at 5.5 per cent is now actually below that of the county (5.9 per cent), having peaked at 11.1 per cent in 1986. The area is set to become the growth area of Leicestershire in the 1990s. The experience of the area of mining decline in Leicestershire shows what can be achieved when problems are anticipated, local authorities act proactively, an overall corporate strategy is prepared and a long-term commitment is made to its implementation, even within the context of limited central-government assistance.

A contrast can be drawn between the impressive success of regeneration in the area of mining decline and the more limited success of regeneration in the other priority areas in Leicestershire – the inner area of Leicester. Obviously, the problems associated with inner-area regeneration in cities are more intractable than that of declining coalfields. Compared with inner Leicester, the area of mining decline has had, and continues to have, a major advantage in having substantial areas of land available for development, not least due to the recycling of land through reclamation. In addition, the area has benefited from major new transport investment, including several bypasses, an upgraded A50 and the construction of the M42. Investment in roads at such a scale would be politically unacceptable in Leicester, even if it were desirable, which it is not. Most importantly, however, the area of mining decline, in contrast to inner Leicester, has had the major advantage of having a regeneration strategy in place for the last 15 years, backed up by strong and clear local political and community support for economic, social and environmental regeneration.

Rural Partnership/Challenge Single Regeneration Budget, RECHAR 11 and City Challenge

In recent years there has been a series of successful bids for additional resources which has added renewed impetus to the regeneration of the area of mining decline. The Leicestershire and South Derbyshire Coalfields Rural Development Area (RDA) was eventually designated in 1993 following a successful presentation to the RDC. A wide range of projects has been implemented by the programme including workspace, community-support initiatives and environmental improvements often related to the National Forest. Among its many achievements, however, has been the fostering of a greater awareness of the area's potential (the programme has a high profile amongst local communities) and the implementation of many projects, often small in scale and local in impact, which otherwise would not have happened but cumulatively have had a major beneficial effect.

Designation as an RDA has resulted in projects in the area being potentially eligible to bid for Rural Challenge. A successful bid has been made in the first round of Rural Challenge. The project which won the £1 million prize is the 'Heart of the Forest' Project which involves the redevelopment of a derelict site in Moira, the geographical centre of the National Forest, as the forest's major facility. The scheme includes workshops, training facilities related to the management of the forest and 'green tourism', a tourism/reception/interpretation facility and new offices for the National Forest team. Work is well advanced and is overseen by a partnership led by North-West Leicestershire Borough Council. Other partners include English Partnerships, British Coal Enterprises, the RDC, the National Forest and the county council.

The series of outstanding successes has continued with the area being awarded £3.35 million, as a result of a successful bid to Round One of the SRB Challenge

Fund, and being designated eligible for EU assistance under the provisions of RECHAR 11. The first year of the SRB scheme has been implemented successfully. The lead is taken by the county council in a partnership of local authorities, the Leicestershire TEC and the voluntary sector. In addition, the county council is a partner in three successful Round Three submissions: Loughborough Town Partnership, Leicester Core Area Regeneration and Loughborough University of Technology 'A Competitive Catalyst'. The Loughborough Town Partnership aims to regenerate Loughborough by improving employment and training, especially in the hi-tech sector, helping to redevelop the town centre, providing community transport facilities and employment initiatives aimed at the ethnic-minority population (£1.337 million). Job and new business creation, training, crime reduction, environmental and housing improvements are the aims of the Leicester Core Area Regeneration (£11.754 million), whilst the 'Competitive Catalyst' Project aims to improve the competitiveness of local/regional economies by creating extensive business support for existing and new hi-tech firms, so that research projects can be transferred into the market place (£0.1 million). Finally, the county council is a partner in the £37.5 million Leicester City Challenge Programme with responsibility for seventeen projects covering most aspects of county council service delivery. In all these programmes, the Planning and Transportation Department co-ordinates the county council's input.

Rural Strategy and Action Programme, Local Agenda 21, Environment City and Environs

Initiated jointly by the county council, the RDC and the Leicestershire Rural Community Council, and set within the context of the structure plan, the Rural Strategy and Action Programme is an attempt to find a jointly shared and agreed sustainable vision of the future for the rural areas, whilst recognizing the often-conflicting pressures for change. It was prepared by over fifty agencies or individuals, and co-ordinated by the county council. The action programme contains over 170 actions or projects grouped under the headings of economy and tourism, community facilities and activities, community care, education and training, housing, conservation and the environment, transport and traffic and access to information and advice. If more resources become available for Leicestershire's rural areas, then implementation of the programme will be accelerated.

Local Agenda 21 action plans for improving the quality of life and the environment in Leicestershire are prepared under the direction of the Forum for a Better Leicestershire, chaired by Professor Mike Brown, Pro Vice Chancellor of De Montfort University. The county council plays a leading role in the work of the forum, helping to co-ordinate and drive the process forward. The council also provides financial support for the forum directly and via a service agreement with Environ (the agency charged with representing the 'green' interest). Consultation with countryside groups and particularly those that are under-represented in the decision-making process, is a key step leading to the preparation of action plans and has recently been completed. The resulting report of consultations will form the basis of the preparation of targets and proposals for the Leicestershire Local Agenda 21. Among the topics which have been debated during the consultative process are the development of green businesses, local purchasing, the promotion of environment-friendly farming, environmental auditing, waste minimization and recycling.

Particularly important is the designation of Leicester as an 'environment city'. The Environment City Campaign was founded in the 1980s in the belief that action could be stimulated by designating a small number of cities as working models of sustainable development. Leicester was the first of the four environment cities to be established in Britain, where commitment to partnership has led to a new emphasis on placing all partners on an equal footing, and encouraging them to work together towards common goals. The county council was a member of the consortium that put together Leicester's bid for the environment-city designation and has been fully involved in the initiative since its launch in 1990. Support for environment-city status from the county council has taken the form of member and officer commitment and involvement, partnership working and financial support.

A recent environment-city initiative is the commissioning of the Sustainable Jobs Project, which seeks to support the creation of new opportunities for paid work which contribute to sustainable development in and around Leicester.

Economic initiatives

Within the Planning and Transportation Department, the Economic Development Unit (EDU) reports directly to the Assistant Director responsible for strategic planning as well as economic development. In addition, the Training Services Business Unit – which is the largest supplier of training in Leicestershire – reports to the Chief Planner who has overall responsibility for all three activities. Economic development and training activities inform the structure plan preparation process and, in turn, are guided by the approved plan. As well as having the opportunity of commenting on major planning applications, the EDU is able to make its contribution to the debate on, for example, the emphasis to be given to inward investment and the amount of new employment land to be released, compared with measures to assist existing industry. As required by statute, an annual Economic Development Action Plan is prepared and subjected to consultation. The plan is prepared within the context set by the structure plan as is the Tourism Strategy and Action Programme. In order to encourage the formation and growth of ethnic-minority businesses, the EDU has also prepared a Minority Ethnic Business Development Strategy and Action Programme, the first of its type prepared in the UK.

Following consultation on the Economic Development Action Plan, an economic partnership was formed in 1993 comprising the Leicestershire TEC, the Chamber of Commerce, the universities, representatives of the district councils, the county council and a representative of Business Link. Working groups have been established on infrastructure, employment, training, business support and knowledge transfer. A county-wide economic development strategy and operation plan has been prepared. The structure plan, regional planning guidance and the TPP have all guided the preparation of this document to which all the main players in economic development have ownership.

The EDU continues to attract an impressive list of inward investment into Leicestershire, particularly into the area of mining decline. In accordance with the structure plan, a science park is being built close to the centre of Leicester as part of the City Challenge Programme. A similar facility beside Loughborough University, which was proposed and initiated by the previous structure plan, is being expanded. An innovation/technology centre, funded by City Challenge, has been built and launched in partnership with De Montfort University. Sites for three of the four proposed HQESs have been identified as has a site for the major distribution facility close to the East Midlands Airport. The latter involves recycled land and has rail access.

A shopping strategy has been prepared for the greater Leicester area in partnership with the relevant district councils. This translates the broad structure plan strategy into more detailed policies and proposals including the identification of specific sites to meet retailing requirements to 2006.

Waste Disposal Management Plan/Waste Minimization

The Waste Disposal Management Plan, prepared under the 1990 Environment Protection Act, establishes the present and future pattern of waste arising and puts forward a preferred broad-brush strategy to deal with them. The Waste Local Plan, prepared under planning law, uses this information on future needs to guide the formulation of proposals and policies required to deliver the Waste Disposal Management Plan in land-use terms.

The county council's Waste Disposal Management Plan was adopted by the county council in November 1995. In order of priority, the strategy places emphasis on waste minimization, reuse, recycling, composting, anaerobic digestion with the residue going to landfill rather than incineration. At a maximum, the volume of arisings to be landfilled could, by 2006, be reduced by 38 per cent compared with 1994 levels. A number of partnership initiatives with industry are being pursued in order to minimize waste, increase reuse and recycling (including measures to bring about greater use of secondary aggregates in new construction), increase the use of washable rather than disposable nappies and reduce the use of packaging. Of particular interest is the East Midland Waste Minimization Initiative. Funded by six project partners, the pilot initiative's purpose is to show commerce and industry how significant savings on cost, as well as less damage to the environment, could be achieved through better waste management. Ten Leicestershire companies, representative of the local industrial structure, were invited to partake. They received a waste audit and guidance on implementing waste minimization. The potential financial savings identified through the audits on all ten companies totalled £3 million. Just as important as the waste costs were the potential environmental savings. It has been estimated that the initiative would contribute to a 10 per cent reduction in water use and effluent, about a 50 per cent reduction in air emissions and about a 50 per cent reduction in landfill.

Assessment and conclusion

A strong strategic planning approach has been adopted by Leicestershire County Council with the structure plan being the central strategic document under-pinning many more detailed plans and action programmes. The relative weight to be given to economic development compared with environmental regeneration, and how the environmental impact of economic development can be reduced to acceptable levels, are addressed by the plan. The debate about how conflict can be resolved between economy and environment, which is a technical and political debate, is conducted openly resulting in a long-term strategy for the future of Leicestershire which all the main stakeholders can sign up to or at least accept.

Undoubtedly, the strong strategic and holistic approach has been a factor in winning an impressive number of competitive bids for resources to regenerate the two priority areas: the area of mining decline and the inner area of Leicester. The regeneration of the former area has been so impressive that its priority-area status may be reconsidered in the next review of the structure plan.

Progress on implementing the transport choice has been better than could be expected. Land for development is generally being released in locations offering, or the potential to offer, transport choice. The TPP has been re-cast in a way which assists implementation of the structure plan. Progress is being made on improving bus-services, cycleways, bus-priority measures, traffic calming and the movement of freight from road to rail. There is, however, no room for complacency. Usage of the first phase of the Ivanhoe Line has been disappoint-ing. The success of transport choice depends on buses becoming a much more acceptable and desirable form of transport: this remains a matter of hope rather than expectation. Regeneration of Leicester and the county towns continues apace assisted by mechanisms such as City Challenge and the SRB. A major issue which remains unresolved is how to protect and enhance the economic viability of city and town centres whilst improving their access by public transport and possibly reducing access by car.

In Leicestershire the main concerns, as expressed by politicians, relating to the interface between economic development and environmental conservation, arise from the growth of car traffic generally, the growth of HGV movements (Leicestershire is an attractive location for distribution activities), the impact in terms of noise, potential wind and water-borne pollution and HGV movements arising from the mining and working of minerals and waste treatment (especially landfill) and the visual impact of large distribution sheds.

The growth of car and HGV traffic is being addressed through implementation of the Transport Choice Strategy. Reducing the impact of minerals and waste disposal activities to acceptable levels is being pursued through the 'plan-led' approach, the review of old minerals permissions and working closely with the new Environment Agency. Measures are being taken to locate new distribution activities on sites which are serviced, or can be serviced, by rail. In addition, the high-quality employment sites proposed by the structure plan are to be protected for B1 use (especially research and development and light industry), rather than B8 (warehousing). This restriction has been supported on appeal.

Using mechanisms such as Local Agenda 21, environment city and Environ, initiatives such as waste minimization, research on how to define, foster and attract 'green' industries are being pursued in order to capitalize on Leicester's first environment-city designation.

References

Johnston, A. (1988) 'Key decisions on out-of-town shopping centres', *Association of County Councils Gazette*, November.
Sabey, P. (1988) 'Shopping: The challenge of the 1990s', *Association of County Councils Gazette*, November.

11 Strategic environmental assessment and local land-use planning
The case of Mendip

Jo Milling

Introduction

In the UK, according to planning-policy guidance notes, the planning system should seek to 'enable the provision of homes and buildings, investment and jobs in a way which is consistent with the principles of sustainable development' (PPG1). A key component of the planning system is the development plan which 'provides an essential framework for planning decisions and should convey a clear understanding of the weight to be given to different aspects of the public interest in the use of land' (PPG 12).

In areas of the UK where a two-tier local authority system is in operation, the development plan is made up of a structure plan produced by the county council and a local plan produced by the district council. The structure plan sets the overall policy framework and looks strategically at development needs. The local plan interprets this guidance locally, by defining sites for development and by formulating locally relevant policies. In unitary authorities, both stages form part of the Unitary Development Plan. Clearly the local plan has a role in establishing a local authority's approach to promoting economic development and protecting the environment. These two aspects of planning policy are often seen as being in conflict. However, it may be more useful to talk in terms of integrating any local authority's approach to economic development and the environment. For example, in a district such as Mendip there is a strong relationship between the health of the economy and the environment. In many cases it is the quality of the district's environment, which attracts new businesses into the area or persuades existing businesses to stay.

The local plan is also closely linked with the council's other corporate strategy statements, the economic development and environment strategies being of obvious importance. The corporate approach to community development (particularly via Local Agenda 21), housing issues, rural issues and leisure provision also have a bearing on the local plan. In many cases the local plan is the vehicle by which the council's corporate aspirations are translated into tangible policies and proposals. The local plan therefore has a central role in balancing and integrating the economic, environmental and social implications of land-use.

Strategic environmental assessment as a tool to promote sustainability

The concept of sustainable development lies at the heart of Mendip's approach to its local plan. There are many diverse definitions of sustainable development, the well known Brundtland definition is a good starting point (WCED 1987). However, the definitions put forward by Jacobs (1991) more closely reflect the stance taken in producing the Mendip District Local Plan. For example, Jacobs writes that:

> The environment should be protected in such a condition and to such a degree that environmental capacities (the ability of the environment to perform its various functions) are maintained over time – at least at levels sufficient to avoid future catastrophe, and at most at levels which give future generations the opportunity to enjoy an equal measure of environmental consumption.

Strategic environmental assessment (SEA) is one means of ensuring that policies, plans and programmes contribute to a move toward sustainability. It ensures that all aspects of the environment are systematically considered and that decisions are taken with the benefit of a full understanding of their environmental consequences. The environmental assessment of a strategic document, such as a local plan, allows the overall strategy for development in an area to be assessed with all the alternatives considered as part of the assessment process. The local plan sets the context within which site-specific projects and proposals will be worked up. These may, in turn, be subject to project-focused environmental-impact assessment (EIA).

The Mendip experience with SEA

This section describes the approach which has been taken at Mendip District Council to the preparation of a local plan. It shows how SEA has been used to apply the principles of sustainable development to plan-making.

A profile of Mendip District

Mendip is an essentially rural district in the north-east corner of Somerset. It has five small towns which form a vital and integral part of a long-established landscape. The area is facing major but often incremental changes from external pressures.

The environment of Mendip is generally of high quality. The Mendip Hills are a nationally important area of outstanding natural beauty and the Somerset Levels and Moors are an area of internationally important wetland habitat. The district has an exceptionally rich heritage of historic sites, structures and buildings. These include the outstanding ecclesiastical sites of Glastonbury and Wells, the cave sites, burial mounds and henges of the Mendip Hills, a large number of Roman settlements and roads, a range of industrial archaeology, ancient field systems and tracks, historic parks and a variety of historic buildings.

The district economy has a strong agricultural base, providing 5% of the area's jobs in 1991, three times the national average. However, continuing changes in the agricultural industry, the need to diversify and the often-unmanaged conflict of interests between countryside users are all placing pressures on the Mendip environment. The manufacturing sector is a major employer, accounting for about one-third of employment in Mendip, compared with one-fifth nationally. Despite the importance of agriculture and manufacturing within the district, the service sector is a growing component of the economy. Tourism in particular generates a range of economic activity and employment opportunities, many in rural locations where other job opportunities are limited.

The Mendip District Local Plan

Mendip District Council began the process of preparing a district-wide local plan in 1993 to replace existing sub-area plans. A commitment was made at that time to ensure that all aspects of the council's work incorporated the principles of sustainable development. The council therefore decided to carry out a SEA of the new local plan as it was prepared.

It was apparent that if the environmental assessment was to be truly part of the process of plan preparation the council needed to adopt a more rigorous methodology than had been previously applied. This had to include clear stages, with the formulation of strategy followed by formulation of objectives, proposals and policies. Each stage needed to be environmentally appraised and the whole process needed to be iterative. It was also important that the methodology made provision for full and early public participation with real opportunities for local communities to shape the plan.

Methodology for plan preparation

The methodology drawn up at this time is summarized in Figure 4. The first stage of the process was to draw up 'environmental stock criteria'. These identify elements of the environment which contribute to its overall quality and which are valued by local communities. The Mendip Environment Forum, technical consultees, local interest groups and parish councils were all involved in identifying these criteria. At the same time, work was started on the plan's strategy. The strategy and criteria developed a symbiotic relationship and the strategy was formulated to take the emerging picture of the 'environmental stock' and its components into account. The preparation of a state-of-the-environment report helped a great deal in this process by providing a comprehensive list of the available environmental data. Much of this data had not previously been seen to be relevant to the plan-making process.

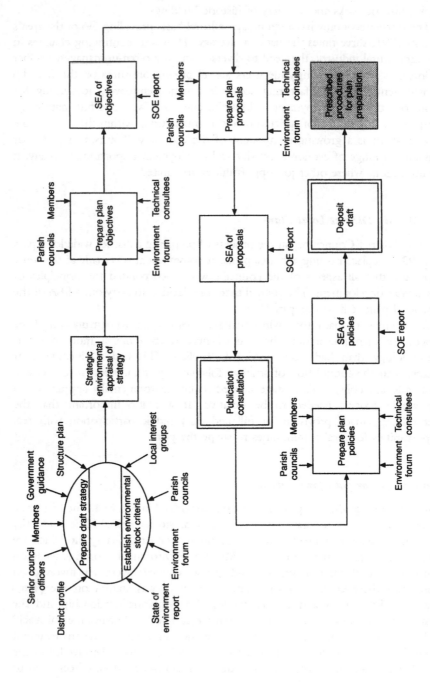

Figure 4 Methodology for plan preparation including strategic environmental appraisal

Once the environmental stock criteria had been established and the plan strategy formulated, the SEA was carried out. More-detailed objectives for the plan were also formulated following on from the objectives of the overall strategy and these were appraised at the same time.

The strategic environmental assessment process

The process of environmental assessment involved making an informed judgement about the likely impact of a strategic aim or objective on each valued aspect of the environment. It sought to ensure that the impacts of all the proposed activities on each of the environmental stock criteria were systematically considered. The environmental stock criteria that were used are set out in Table 13.

The appraisal exercise was undertaken by a multidisciplinary panel, including senior officers from across the council and specialists from the Somerset Wildlife Trust, the Environment Agency and the Countryside Commission. The impact of each strategic aim and objective on each environmental stock criterium was assessed in turn. Each impact was discussed by the group and recorded on the impact matrix. Impacts could be positive, negative, unknown due to lack of information or unknown because the outcome depended on decisions that were to be taken at a less strategic level in the plan hierarchy. Most importantly, an explanation of the judgements made was recorded on a commentary matrix. Judgements were based on the best information available and on the professional expertise of the officers involved.

The process was designed to be as transparent as possible but not to involve excessive time and research. It ensured that each of the environmental stock criteria was considered, and it recorded those areas where information was lacking or where the impact would depend on the way in which the strategy or objective was implemented.

The process was designed to be iterative as it created the opportunity to go back and alter the strategy and objectives where negative environmental impacts were identified. From the outset the process did not include economic or social factors as stock criteria, instead it looked purely at the environmental consequences of the strategy or objective. It was accepted that a negative environmental impact could be accepted in order to secure economic or social gains and consequently that in some instances the environmental assessment would merely make the environmental impacts of a particular decision explicit.

Having drawn up, assessed and agreed a strategy and objectives for the plan, the process moved on to look at proposed sites for development. The same methodology was applied and the likely impact of development being allowed on a series of sites was assessed against the various environmental stock criteria. Again, this was done with the best information available and made use of the professional judgement of a multi-disciplinary panel.

Table 13 Environmental stock criteria used in the environmental assessment process

Global sustainability

Transport-energy efficiency – trips. Some pollutants are affecting the global atmosphere and potentially causing climatic change, reductions in the ozone layer and acid rain. These problems are caused by gases such as carbon dioxide, nitrous oxides and sulphur dioxide. One of the main sources of these pollutants in Mendip District is vehicle emissions from fossil fuels. Reductions in the need to make trips and in the length of those trips will therefore help to reduce global atmospheric degradation.

Transport-energy efficiency – modes. The use of modes of transport which produce fewer harmful emissions for each mile travelled by one person than the private car will help in reducing global atmospheric problems, as set out above. Less damaging means of travel include walking, cycling and public transport.

Built-environment energy efficiency. Minimizing the use of energy in buildings will contribute to reducing global atmospheric problems. This might include siting and landscaping buildings in an energy-efficient manner as well as energy-efficient design and specification of buildings.

Renewable-energy potential: The substitution of energy generated from renewable sources for that generated from burning fossil fuels will help to reduce global atmospheric problems. There is some, limited, potential for renewable generation in Mendip.

Rate of CO_2 fixing. CO_2 is one of the main gases affecting global atmospheric systems. It is released to the atmosphere when fossil fuels are burnt but can be 'fixed' in the form of plant tissue as vegetation transpires. Trees are amongst the most effective 'fixers' and it is important to protect and plant them.

Wildlife habitats. Important habitats in Mendip include woodlands, grasslands and wetlands. Much wildlife interest is also found in hedgerows. Many sites are designated as having special wildlife value but the wider countryside is also important for many species, and in linking areas of more specialized habitat. Important sites may also be found within the district's towns and villages.

Natural resources

Air quality. As well as affecting the global atmosphere pollutants have an effect locally. In addition to those mentioned above, dust, radon, methane and CFCs are of concern.

Water conservation: High rainfall, the occurrence of water-bearing rocks and the watershed of the Mendip Hills make Mendip strategically important for water issues. The district contains headwaters of rivers flowing north and source waters of internationally important wetlands in the Somerset Levels and Moors. The conservation of water to feed these important watercourses is therefore important.

Water quality. Important aquifers underlie large areas of Mendip. Water quality is generally high but is vulnerable to pollution. Similarly, water quality in the district's rivers is generally high. Both ground and surface waters are susceptible to pollution from sewage, farm slurry and chemicals, contaminated surface water and trade effluent.

Conservation of undeveloped land: Land is a vital resource for agriculture and wildlife. Once built on it is extremely unlikely that it will be returned to such 'green' uses.

Mendip has a high proportion of grade 3 agricultural land but is relatively short of the best grades (grades 1 and 2) land. Dereliction of land is also of concern, whether in the towns or the countryside.

Soil conservation: Conservation of the soil itself is an important concern since it is a vital resource for agriculture and wildlife and is essential in determining water quality. Issues include retention of the soil and contamination.

Minerals conservation: Mendip is rich in minerals, principally limestone and peat. The extraction of minerals is not within the remit of this plan, but rather is dealt with by the Minerals Local Plan being prepared by Somerset County Council. However, some of the impacts of mineral extraction and more broadly, developments, which result in demand for aggregates, may fall within the remit of the plan.

Local environmental quality

Landscape character: Mendip has a diverse and high-quality landscape. The character of the area depends not only on its physical attributes, such as geology, soil, topography and vegetation, but also on the uses to which the land has been put, both now and in the past, patterns of building and settlement and the management of the land. The quality of the landscape is largely determined by its visual characteristics, but noise, smell, historical associations and cultural nuance are also important in determining how the landscape is perceived.

Built-environment 'liveability': The quality of the built environment is of great importance in determining the quality of life of those who live in towns and villages. The juxtaposition of different types of development and convenient access to homes, jobs and services are vital to providing a convenient and pleasant living environment. The protection of the historic character of Mendip's towns and villages, provision of open space and urban design are also important.

Cultural heritage: The district's history is essential to its sense of cultural identity and lends many parts of the area a distinct sense of place. Its maintenance is important to the well-being of the local community, as well as for its own sake and its educational value. The district's stock of environmental cultural assets includes historic buildings and streets, the historic landscape including some parks and gardens and archaeological sites. Geological exposures which are important for their educational and scientific value are also included here.

Public access to open space: Access to open space is important in determining people's quality of life. This might include access to open spaces in towns but also access to the countryside for informal recreation.

Building quality: The maintenance of existing buildings and the quality of the design and construction of new buildings are important factors in ensuring a high-quality living environment.

Source: Based on Department of the Environment (1993).

The benefits of SEA as an external checking mechanism

The approach to SEA that was adopted was relatively quick, it introduced areas of environmental concern not previously considered, it was accessible to all and it made explicit the environmental choices being made. However, it is important to recognize that the results recorded in the matrices were arrived at through reasoned judgement and discussion and as such the individual assessments were neither objective nor quantifiable. It should also be acknowledged that the presentation of the results in the matrix format could have been misleading if it was taken to imply a scientific basis, which did not exist. Nonetheless, the commentary matrices were invaluable in explaining how decisions were arrived at and in pointing to further research that needed to be done. They also indicated how some proposals needed to be implemented if environmental impacts were to be avoided or reduced. Perhaps most importantly, these explanatory matrices made the decision-making process much more transparent and amenable to external scrutiny.

This type of SEA is largely external to the plan-formulation process. The assessment is designed to check the consequences of a previously drafted strategy, objective or site proposal. However, in this case a close working relationship did develop between those involved in the drafting of the plan and those concerned with the assessment process. Indeed, as the values expressed in the preparation of environmental stock criteria and the subsequent assessment were built into the initial draft of the plan's strategy, objectives and proposed sites, the need for an external checking mechanism was in some ways reduced.

The environmental capacity approach

Further work on the plan presented the opportunity to integrate environmental concerns into the plan-making process more fully as an alternative to using SEA as an external checking mechanism. The first draft of the strategy, objectives and site proposals formed the basis of a consultation paper which was published for public comment in 1995. Initially, this process was to work towards an end-date of 2001; however, because of public consultation, it was decided to extend this end-date to 2011 and to revise the plan's strategy with this in mind.

This coincided with a review of the Somerset Structure Plan, which also works to an end-date of 2011. The structure plan interprets regional guidance and indicates (amongst other things) the extent of new development required in each district, particularly the number of new houses to be built. Numbers are based on a combination of estimates of new household formation from the existing population and migration to the area. The local plan must be in general conformity with the structure plan before it can be adopted as the formal policy of the council. Where the plans are based on different principles or employ different techniques this can lead to conflict. In this case, the structure plan uses a basic checklist approach to SEA. The structure plan may also, in its turn, come into conflict with central government's regional guidance which, for example, proposes the aggregate number of new houses needed in the county as a whole.

The re-evaluation of the local plan strategy provided the opportunity to incorporate the principles of environmental assessment into plan formulation in a more structured way. The environmental stock criteria were used as a framework for researching the spatial pattern of development across the district.

This analysis was then used to identify those areas where development would be most sustainably located. This exercise formed the first step in an assessment of environmental capacity which subsequently tested the district's ability to accommodate the level of development suggested by the structure plan.

The concept of environmental capacity is widely used in ecology to indicate the ability of an ecosystem to support a species or activity without losing the ability to renew itself. It is often also applied to an ecosystem's ability to support human activity. When applied to the global scale, this concept is an integral part of the concept of sustainable development. However, environmental capacity is also a useful concept when interpreting and applying sustainable development at the local level.

Local-authority boundaries do not, of course, define ecosystems. However, it has been convenient to look at the district environment as one system whilst recognizing the linkages at its boundaries. It has then been possible to analyse the spatial patterns at work within that system and to explore the ways in which social and economic activities affect the environment's ability to sustain itself.

Table 14 shows the environmental, social and economic factors which have been investigated and the indicators which have been used at the district level. The environmental factors are the same as those used as environmental stock criteria in the earlier SEA with the exception of those which relate only to local environmental quality. The factors used are intended to highlight issues relating to the earth's long-term ability to renew itself. Issues of local quality of life are addressed later in the process.

Many of the indicators used have environmental, social and economic dimensions and this highlights the interrelated nature of these aspects of land-use planning. For example, poor environmental conditions, social problems and economic downturn will often be found together. Similarly, environmental problems will often be expressed as social or economic concerns and *vice versa*. Any land-use strategy, which hopes to address sustainable development must address all three sides of the issues.

The capacity of Mendip's towns for development

Having established a strategy for land use on a district-wide basis, the process moved on to look in more detail at the ability of the chosen locations to accept development. This process was again iterative and it was accepted that the strategy would be revised if it were found that the chosen locations could not sustainably accept development.

In this case, the areas identified as being most suitable to accept development were the district's five towns. Conditions in each town were researched in more detail to establish their capacity to accept growth. Again, a range of environ-
mental, social and economic factors was investigated, including those relating to local environmental quality. This exercise sought to identify those areas which could be developed without loss of environmental quality and those which would most readily provide opportunities to live more sustainable lifestyles.

Table 14 Environmental, social and economic factors researched and the indicators used

Environmental factors

Transport-energy efficiency – trip generation
- Travel to work patterns
- Job/housing ratios
- Retail provision

Transport-energy efficiency – modes
- Potential for walking and cycling
- Public transport analysis

Built-environment energy efficiency
- Topography
- Density of development

Renewable-energy potential
- Assessment of viable locations for energy generation using wind turbines

Rate of CO_2 fixing
- Mapping of woodlands

Wildlife habitats
- Mapping of designated sites
- Mapping of areas designated in the Local Biodiversity Action Plan as having value or potential

Air quality
- Mapping of areas affected by radon (it was recognized that other air-quality issues needed further investigation and that information was not currently available)

Water conservation
- Analysis of availability of public water supply (Local Catchment Management Plan did not identify any major threat to the supply of water within the remit of the local plan)

Water quality
- Mapping of ground-water-source protection areas identified by the environment agency
- Mapping of flood-risk areas identified by the environment agency

Conservation of undeveloped land
- Assessment of the potential to reuse brownfield sites within the built-up area

Soil conservation
- Agricultural land classification

Minerals conservation
- Analysis of spare capacity within the existing infrastructure

Social and economic factors

Unemployment
- Unemployment rates
- Numbers of individuals unemployed

Access to local jobs
- Job/housing ratios
- Travel-to-work patterns

Housing need
- Analysis of numbers of households requiring affordable housing

Housing demand
- Analysis of population change

Accessibility
- Public transport analysis
- Potential for cycling and walking

Access to social facilities
- Distribution of healthcare, education, leisure and other facilities

Source: Based on Barton *et al.* (1995).

The concepts of critical natural capital and constant natural assets were relevant to this approach but were not explicitly employed. Rather, the implications of development in specific areas were assessed and decisions on the appropriateness of different development options were made in each case. This allowed the complexity of local conditions to be considered more sensitively without losing sight of the effect of a specific decision on the full range of environmental stock.

The approach described so far provided a picture of the amount of development which could be accommodated within each town without exceeding its environmental capacity. It was also necessary to explore the need for growth and the desire of the local community for development. The need for growth can be investigated through a purely technical analysis of population, employment and social trends. Getting a well-balanced and thoughtful picture of the local community's attitude to development is more difficult. Public awareness of the planning system and the issues involved must be increased before a consensus on the community's future can start to emerge. The local view on the future of each town can then be used alongside the work on environmental capacity.

Benefits of the environmental capacity approach

The application of this methodology, which required a substantial commitment on behalf of the local authority, made the principles of sustainable development central to the process of plan formulation. It allowed for the integration of environmental, social and economic concerns and facilitated an attempt to identify the ways in which the needs of all three areas can be balanced. Where significant environmental losses were associated with the realization of social or economic objectives alternative approaches were sought. Where conflicts remained, the basis for decision-making was made transparent. This approach helped to ensure that the land-use strategy explicitly considered the economic, the social and the environmental dimensions of sustainable development. It also enabled the need for development in an area to be considered alongside an assessment of the ability of that area to accept development.

Thus, this approach helped to make the impacts of decisions on land-use planning more explicit and open to external scrutiny. While this process has had a range of benefits, it should also be noted that in some instances it has produced real conflict between the need to promote social and economic development and the desire to support environmental protection. In Mendip, this conflict has been based particularly on the desire to protect the countryside and the need for greenfield sites to be released for development if the demand for new homes is to be met until 2011.

Conclusions

It has become readily apparent in recent years that the planning process needs to pursue environmental, social and economic concerns in an integrated and balanced way if sustainable development is going to be promoted more effectively. The SEA of development plans provides a useful means of introducing environmental concerns and issues which may not otherwise have been seen as being an integral part of the planning process. SEA ensures the systematic consideration of the full range of environmental issues and highlights the environmental consequences of a proposed policy. The systematic and explicit identification and analysis of those elements of the environment that are valued by the local community is also invaluable in heightening the level of priority awarded to the environment and in raising awareness of the threats to the environment. However, SEA does not remove the need for hard political decisions to be taken where environmental concerns come into conflict with economic or social considerations or where choices between different areas of environmental concern are raised.

The concept of environmental capacity takes a further step towards integrating the principles of sustainable development into the planning process. The concept requires that environmental, economic and social issues are seen as parts of a system rather than as separate or conflicting concerns. Such a holistic view will enable a more integrated approach to development and land-use planning. Greater familiarity with the concept of environmental capacity and its implications can also highlight the inextricable links between the land-use planning system and the multitude of other issues that are addressed by other corporate strategies within the local authority and the strategies of other organizations.

In Mendip, the use of these concepts has altered the way in which the local plan has been produced and the strategy which has resulted from this process of plan formulation. The research that has been used to inform the development of the plan has been more broadly based than previous local plans with a wider range of environmental, social and economic issues being investigated. Public participation has been more meaningful and this alone has raised awareness of the environmental issues to be addressed and the needs and wishes of the community. Consequently, the land-use decisions that are embodied in the plan can be both more environmentally sensitive and more responsive to the aspirations of the community.

The research carried out at this early stage in the process will also mean that design solutions can be sought to resolve or at least reduce conflicts on specific sites. In some instances it has, however, been difficult to meet environmental, social and economic concerns without generating conflict. This has been particularly marked where the environmental and the socio-economic needs of the towns seem at first glance to be mutually exclusive. In this situation, new and potentially more radical solutions must be sought.

References

Barton, H., Davis, G. and Guise, R. (1995) *Sustainable Settlements: A Guide For Planners, Designers and Developers*, Luton: UWE/LGMB.

Department of the Environment (1993) *Environmental Appraisal of Development Plans: A Good Practice Guide*, HMSO.

Jacobs, M. (1991) *The Green Economy*, London: Pluto Press.

WCED (World Commission on Environment and Development) (1987) *Our Common Future*, Oxford: Oxford University Press (Bruntdland Report).

12 Strategic planning for the sustainable city-region

The case of Greater Manchester

Joe Ravetz

Introduction

This chapter examines the longer-term prospects for integrating environmental management with economic development. It is based on a detailed case study of an integrated sustainable development strategy that has been developed as part of the Town and Country Planning Association's 'City-Region 2020' project for Greater Manchester (Ravetz 1996; Ravetz *et al.* 1999). This project sought to establish whether a city-region could achieve environmental sustainability within the time-span of one generation, taken here to be 25 years. The investigation examined environmental, economic and social trends and dynamics, it applied sustainability principles to each and it set out policies and actions to promote a transition toward sustainability. However, these policies and actions were not proposed as final solutions but as ways of exploring the problems that may be associated with such a transition (Breheny and Rookwood 1993).

Given its focus on long-term strategic change, this chapter examines three key issues: first, it examines the parameters that may be used to assess the degree of environmental sustainability in a local economy; second, it considers the extent to which environmental management and economic development have been integrated to date, and how far is there to go; and third, it assesses the role of local government in achieving such integration.

Environmental sustainability in a local economy

While the broad principles of sustainability may be clear, their application to real-life problems is very problematic. This is one reason why many businesses have little time for it, and why currently the degree of integration of economic and environmental policies in local authorities is at such a low level (Gibbs *et al.* 1996). One problem is that there is no single measure of the sustainability of a firm, a sector or a local economy. Another is the interdependence of activity at every level – with difficulties in defining environmental capacities, let alone economic boundaries. A third issue is that sustainability is both a technical issue and an ethical concept surrounded with value-judgements. Single solutions cannot cover such a complex question. In this section several key parameters for environmental sustainability in a local economy are discussed, as different perspectives of the economy–environment linkage:

- integrated assessment – a multi-dimensional framework for complex systems;
- stocks and flows – environment-economic capital and welfare;
- limits – capacities and thresholds in environmental resources;
- dynamics – environmental strategy as a creative evolution in firms and sectors;
- linkages – relationship of environmental to other dimensions of sustainability;
- transitions – ecological modernization in the context of wider trends and dynamics.

An integrated systems approach

Any industrial sector can be seen as a system; that is, as a set of inter-connected components and activities which constantly interact to maintain the viability of the whole (Clayton and Radcliffe 1996). As with any system, the whole is greater than the sum of the parts, and while new patterns may organize themselves, its survival depends on feedback for self-regulation and containment of pressures or externalities on other systems (Portugali 1997; Bossel 1996). Looking beyond a single industry at the environmental sustainability of a local economy, the system under consideration is much wider – within the boundaries of the system are cultural, social, political, economic, technological, spatial and environmental factors. By listing the key factors and exploring possible linkages or tensions between them it is possible to build up a framework for integrated assessment (Ravetz 1999b). Figure 5 depicts some of the factors that need to be considered within such an integrated assessment as applied to the process of eco-modernization.

Such an integrated assessment can highlight any linkages that are relevant to the discussion. For instance, for the objective of environmental protection, linkages can be traced to efficiency and innovation; GDP and competitiveness; levies and regulations; and employee and stakeholder pressure. For the objective of competitiveness, linkages can be traced to consumer demand and labour skills, to innovation and management systems and to environmental quality and resources. The range of factors involved, and the complexity of their interaction, shows the significance of the holistic context for any one problem or solution. It can also be used to chart the various parameters of stocks, limits and dynamics.

Stocks and flows

One approach to integrated assessment is to look at the totality of stocks from an economic welfare perspective (Nijkamp *et al.* 1992). For a firm, traditional balance sheets can be extended to a triple bottom line which includes environmental and social as well as economic factors (Elkington 1997). Likewise, for a city or region, the general goal is to maximize the sum of social, economic and environmental capital assets and human welfare, assuming that these can be

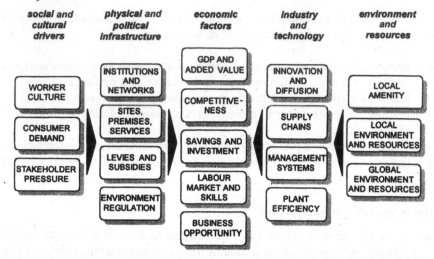

Figure 5 Eco-modernization in the local economy. Key factors in the integration of environmental management with economic development: integrated assessment framework (adapted from Ravetz 1999).

aggregated. Between communities or regions, transfers of materials, assimilative capacity and human capital may be less or more sustainable, depending on the relative resources, capacities and development paths of the regions (White and Whitney 1992). Between generations, there are trade-offs between present and future costs and benefits, depending on assumptions for futurity or the social rate of discount.

In practice, the territorial model of discrete spatial units is shifting to a functional model of multiple specialist nodes within a wider economic order (Friedman and Weaver 1979). Therefore, the welfare balance sheet for an individual city-region is meaningful only in relation to the dynamics of the global system. In practice, such a balance sheet contains many different and incommensurable values. Accounting systems that attempt to respond to this situation, such as the index of sustainable economic welfare, depend on many subjective weightings (Jackson and Marks 1994), whilst many indicators of value are based on cultural perceptions of trust and risk (McNaghten *et al.* 1995). The result is that, just as in a firm, a local/regional balance sheet approach is one parameter rather than an overall assessment.

Environmental capacities and limits

The next step is to consider environmental capacities or thresholds for the physical metabolism and impact of the local economy. These capacities apply at local and global scales, for natural and human systems, and in the present and future. Thus, there are several kinds of capacity, from the neighbourhood to the bio-region:

- spatial or urban capacity – land use, location, access, solid waste;
- resource capacity – depending on how imports and exports are accounted;
- assimilation capacity – pollution, toxics, critical loads;
- ecological capacity – tolerance of species and communities to stress;
- social capacity – tolerance of congestion, noise, amenity, density and statistical risk: each of these involves multiple values and perceptions.

The urban environment capacity approach has so far focused on development and land-use planning, but in practice there are few definitive thresholds (Barton and Bruder 1995). A wider metabolic capacity, that focused on the materials and energy throughputs and emissions in a city-region, would look at the longer-term trends and cumulative impacts of the urban system (Entec 1996). For global capacity, known thresholds and limits have to be aggregated, and assumptions made for local or regional pressures, with principles for distribution of access to resources. Current approaches that are related to broader notions of capacity include the concepts of environmental space (Carley and Spapens 1997), environmental footprints (Rees and Wackernagel 1995) and eco-indicators (Pre' Consultants 1995).

Environment-economy dynamics in firms and sectors

The stocks and limits discussed above are static parameters that need to be complemented with a range of more dynamic measures that reflect the degree of change in business and the economy. It is possible to depict the nature of the industrial response to environmental issues in an evolutionary way shifting from regulatory compliance, to environmental management, to total producer responsibility, to wider stakeholder engagement, to the sustainable firm – the last being the most difficult to define (Wood 1996; James 1994; Welford 1995). Various measures of progress have been proposed; for example, the productive efficiency approach integrates socio-economic and environmental indices and looks at the best-practice frontier as a reference point for actual performance (Tyteca 1996).

To an extent, these firm-level measures can also be applied at the sector level. One approach is with impact-welfare ratios; for example, the UK pilot environmental accounts as depicted in Figure 6 show the ratios of emissions to jobs or added value for each major sector (ONS 1996). If emissions can be linked to environmental capacities and limits, such ratios are a reference point for sectoral strategies or agreements. However, as each sector is embedded in the context of the whole economy, such ratios are only a starting point and a wider whole-economy assessment is needed. Such an assessment could be based on input/ output life-cycle analysis (LCA) or on an assessment based on integrated chain management techniques applied to sectoral material flows (see Wolters *et al.* 1997). These techniques could eventually allow the development of an overall index of environmental performance that could relate to firms, sectors or local economies.

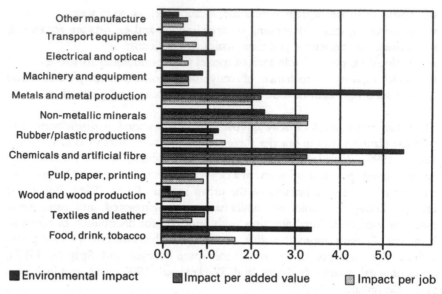

Figure 6 Environmental impact of industry: Aggregated climate change and acid rain impacts by manufacturing sector, with ratio of impacts to employees and added value per sector. Figures apply to percentage of UK totals. Source ECOTEC 1996.

Economic and social linkages of environmental strategy

Naturally, there are concerns about the economic implications of the transition toward sustainability as such a transition may have both positive and negative impacts on other economic or social goals. For each aspect of environmental strategy, local and regional economic impacts can be identified and categorized (if not calculated) as shorter-term adjustment and longer-term adaptation costs (Jacobs 1994). Areas of employment growth could include the protection and creation of environment-related jobs in regulated industries, in sectors that realize increased competitiveness through environmental improvement, in sectors (such as tourism) that are dependent on environmental quality, in the public sector through greater investment in environmental infrastructure and in the environmental industries through growth in demand for their goods and services. By contrast, employment might decline in sectors producing end-of-pipe technologies and in sectors which are vulnerable to the increased costs of environmental control. While the style and flexibility of regulation is a crucial factor for these vulnerable industries, extrapolation from a survey in the Midlands would suggest that up to 30,000 jobs could be at risk in Greater Manchester in industries such as paper, textiles, food, chemicals and distribution (ECOTEC 1991). However, there is also some evidence that the lack of environmental regulation can impair competitiveness in world markets; for example, where higher standards are already enforced elsewhere and where first-mover status is a crucial advantage (DTI/DoE 1994).

It seems likely therefore that tougher environmental regulations could threaten employment in some areas whilst creating jobs in others. However, some European experience suggests that the overall impacts of enhanced regulation are neutral or positive, with accelerated obsolescence in some older industries (Hitchens 1997). If this is the case, then creative and continuously upgraded environmental regulations could stimulate the development of competitive advantage in a way that is compatible with the concepts of ecological modernization.

While much of the current emphasis of industrial environmental management practice focuses on the supply-side of production, a wider urban or regional perspective needs also to address the demand side. 'Factor four' arguments that suggest that it may be possible to double wealth and halve resource use depend as much on the management of demand as on the promotion of new supply-side initiatives (von Weizsacker *et al.* 1997). In this respect, there are obvious examples where demand needs to be influenced more significantly; for example, in relation to transport and energy consumption. In both these instances, demand-side management techniques are becoming common practice, leading amongst other things to the development of new approaches to infrastructure management and provision (Marvin and Graham 1995).

Environment-economy transitions

The transition of economic activity is likely to produce not only quantitative changes in resource usage, but also many qualitative changes in the organization of businesses and local economies. To this end, Elkington (1997) suggests that market transition is likely to be associated with changes in corporate values, with greater emphasis being placed on the trust of consumers, employees and other stakeholders, which will in turn be driven by increases in transparency, with information becoming increasingly available and consumer awareness of key issues increased. Elkington (1997) also argues that new forms of corporate governance will emerge, shifting the emphasis from shareholding to stakeholding forms of accountability, and that partnerships within supply-chains, between competitors, and between industry and its many stakeholders will become more significant. Finally, Elkington (1997) proposes that corporate responsibility will be extended to cover the life-cycle of a technology over a longer (possibly inter-generational) period. The accountability and governance structures within companies and sectors will be connected to a wider set of stakeholders (Lewis 1997). The relationships between local businesses and economies may change in ways that are hard to imagine, but it is possible that the accelerating globalization of business will be complemented by increased localization in environmental and social accountability.

Current environment-economy relations

Greater Manchester, with its population of 2.5 million people, contains some rapidly growing economic sectors in finance, media, education, sports and arts. However, it is also a recognized lagging region with a variety of social, economic and environmental problems (Clark *et al.* 1992; NWRA/NWBLT 1993) (see Table 15).

While they could be further developed, there are some links between economic development and the environment in regional and local policies. For example, at the regional level environmental quality is seen as essential for attracting inward investment, but while the regional economic and planning guidance includes the goal of a green and pleasant region, there is no consideration of the environmental impacts of the regional economy (NWRA/NWBLT 1993; GONW/GOM 1996). At the conurbation level, the *ex ante* assessment of the environmental impacts of the operational programmes for which financial assistance was sought from the European Union Structural Funds included an environmental policy appraisal matrix. However, this included little assessment of long-term trends in sectors such as transport, energy or waste (Environment Agency 1997). Indeed one of five programme headings is concerned with regeneration and the environment, with 17 per cent of the budget, of which the measure 'promoting better environmental practice' is under-spent. Within the planning system, although many environmental policies are included in the local unitary development plans (UDPs), these are often outweighed by the pressures of economic development and regeneration (Gibbs *et al.* 1996). Many issues such as transport and air-quality management need more effective co-ordination at the city-region level.

Within the region, a large number of initiatives have sought to promote improved environmental performance in industry. Some of these initiatives have been successful whilst others have encountered problems. Particularly successful initiatives include the Department of Trade and Industry sponsored Project Catalyst, a partnership that sought to promote the economic and environmental benefits of waste minimization in industry in the region. Other initiatives include those of the Manchester-based Co-operative Bank that has taken a profitable lead with its green and ethical policies and has introduced preferential lending for environmental investment (Co-op Bank 1995). The bank also sponsored the

Table 15 Environment, society and economy in Greater Manchester

- air quality: 200 exceedences of various standards per year
- water quality: largest proportion of 'poor' and 'bad' water systems in the country
- ground quality: 10% of the urban area potentially contaminated or unstable
- waste: 12 million tons of solid waste; 35% of UK hazardous-waste imports
- land use: 10% of the urban area vacant, 6% derelict.
- housing: 35% of all housing unfit, 28% of housing over 100 years old
- population: total out-migration of 25% since 1970; overall density of 45 people per hectare
- health: standard mortality 20% above average
- economy: most indicators 5–10% below national average; total GDP £19 billion
- manufacturing added value per employee: 25% below national average
- largest financial and media centre, largest airport outside London
- education: 20% under attainment; four higher education institiutes (HEIs)

Sources: GMR (1995), AGMA (1994). Comparisons refer to UK average.

National Centre for Business and Ecology in partnership with the four local universities which sought initially to improve the access of SMEs to research and development facilities but is now more focused on initiatives designed to enable action to be taken in SMEs (NCBE 1997). The bank was also instrumental in the North-West Partnership that recently established a pioneering umbrella agency, Sustainability North West.

However, despite the influence of these and other initiatives, SMEs have been slow to respond. This has been shown by a local survey (Hooper and Gibbs 1995) which found that only:

- 10 per cent of (responding) SMEs think they have significant environmental impacts;
- 20 per cent have some form of environmental review or written policy;
- 30 per cent are interested in a green business club, but only 1 per cent attend meetings;
- 35 per cent are aware of environmental management systems and techniques.

These findings are supported by those of a similar national study (Groundwork Foundation 1996). This low-level of interest and awareness cannot be due to the lack of regulation: of the 60,000 businesses in Greater Manchester, about 20 per cent are registered for some kind of environmental regulation, including 31 sites registered for Integrated Pollution Control delivered by the Environment Agency, 5,000 sites for Air Pollution Control delivered by local authorities, 2,000 producers of trade effluent and 3,500 licensed waste producers or carriers (HMIP 1995; GMWRA 1995). A wider sectoral analysis also suggests that perhaps another 40 per cent of businesses have significant material throughput. Meanwhile, it can be broadly estimated that formal environmental management systems have spread to perhaps one in a thousand businesses in the region, and in-house or informal reviews and audits have spread to perhaps one in a hundred companies. In terms of support mechanisms and initiatives designed to promote improved environmental management in industry, one regional survey revealed a plethora of competing and uncoordinated agencies and initiatives which tend to reach the already converted or those keen to placate the regulators (Douglas and Lawson 1998).

Environmentally sustainable economic development

Comparing the key environment-economy parameters as discussed above with the current levels of environmental activity in businesses in the region, several themes emerge for the longer term.

Environmental limits in the city-region

The environmental capacity of the urban system forms the long-term limit for economic development; for a city-region such as Greater Manchester pressures on land, transport and energy highlight this. There is no shortage of land for economic development, with large areas of vacant land and dereliction, but high-quality sites with good amenity and access are pressing on the boundaries of the green belt (AGMA 1994). Long-term trends suggest that employment floorspace per job increases at about 2 per cent per year (Fothergill *et al.* 1987). In Greater Manchester, if this trend continued, employment space requirements would double in 30–40 years, taking up more than the total of vacant/derelict land, and there is already suppressed demand from an ageing and obsolescent building stock. While land use is not yet on the environmental management agenda, it is clear that in the long term economic growth will need to be decoupled from spatial growth.

For transport, large parts of the urban area are already beyond their capacity limits: the 2020 project proposes a stabilization of growth and a shift of 25 per cent of urban road transport to other modes to meet targets for congestion, neighbourhood development, air quality and climate change (RCEP 1994). This is technically achievable, and could well improve the prospects for some sectors of the local economy (Whitelegg 1995). However, it also challenges current trends toward globalization, especially for a location on the European periphery. A co-ordinated approach would aim to integrate supply, demand, infrastructure and technology, extending environmental management to sustainable transport.

Targets for energy and climate change require the application of best practice across all sectors, including the upgrading of all existing buildings, new building to best-practice standards, the application of combined heat and power (CHP) networks, and the adoption of the best-available process technologies in industry. Achieving such targets depends both on national policies and on local actions for integrated longer-term investment in urban infrastructure and regeneration. With the current deregulation and privatization of many basic services and utilities, and falling energy prices, this is difficult; for instance, an ambitious CHP scheme in Trafford Park in Manchester failed to secure long-term commitments from industry (March Consulting 1995).

City-region environmental management framework

Each of these examples hinges on integration, co-ordination and investment – the logical functions of a regional environmental management system (Welford and Gouldson 1993). Such a system or framework in many ways represents best-practice city or regional planning, updated and extended with new agendas in environmental and economic sustainability (Roberts 1994). Such an environmental management framework would contain both reactive and proactive roles:

- reactive monitoring and regulation of environmental conditions, pressures and standards;

- proactive supply-side initiatives in business, technology and human resources;
- proactive demand-side initiatives in infrastructure, market and consumer patterns.

A region is any area where key actors have a common interest, generally coinciding with a political, economic or ecological boundary. With a large and diverse economy, a conurbation such as Greater Manchester is in many ways a viable functional unit for an environmental management framework, although its ten local authorities and many other agencies lack full co-ordination at present (Roberts and Chan 1997). With the new generation of regional development agencies (RDAs) and regional assemblies in England, the regional level is now the focus of attention. In political and economic terms, conurbation-level environmental management is likely to be a subset of a regional framework: while in environmental terms the smaller city-region is perhaps a more appropriate functional unit.

For environmental issues, the general aim is to establish an integrated framework of monitoring, regulation and investment, but at present there are many bodies with fragmented accountability across different media. This reality contrasts with the rhetoric and responsibilities placed on local government by Agenda 21 and many national policies (Richards and Biddick 1994). Bridging this gap will require technical developments such as the application of comprehensive geographical information systems and cumulative assessment and systems modelling. It will also require political moves including stronger enforcement powers, restructuring of public agencies, utility regulation and resources for environmental investment and purchasing policies, whether these come through the RDA, local authorities or other bodies (Sustainability North West 1998).

For economic development issues, the role of local government is that of one amongst many agencies, depending on the future scope of the RDAs, European funds and many others. Local authorities may continue to shift their activities to enabling rather than providing, by co-ordinating many other organizations within a democratic context. In addition, extending this context, new forms of citizen and corporate participation are emerging, following models as diverse as Local Agenda 21 and City Pride.

With such uncertainties and complexities as a background to city-region environmental management, it is perhaps more practical to envisage an open-ended framework for co-operation and co-ordination rather than a single system of direct management control. A city-region environmental management framework would aim to be horizontally integrated between various public, private and quango sectors and industries; vertically integrated between regional, sub-regional and district-level bodies: and environmentally integrated between various physical media and pressures. Local government may be in a good position to take the lead role in such a diverse partnership, depending on how parallel bodies such as the RDAs operate in practice.

Environment-economy opportunities

What would such a city-region environmental management framework do? In general it would take the greening agenda for traditional local economic development towards the goals of environmental and economic sustainability (Gibbs 1994). Typical headings include information, in-house activities, sites and premises, infrastructure and amenity, area regeneration, business-development initiatives, and specialist programmes; and for each of these there is a corresponding environmental management activity (LGMB 1994). This agenda can then be extended in every dimension of the local economy, and it might include total materials management, integrated supply chains, technological innovation, human resources, financial services and consumer awareness. For each of these, local government may have a more direct provider role, or a more indirect enabling and partnership role.

First, such a city-region environmental management framework would co-ordinate comprehensive monitoring, auditing and regulation of the local environment and material metabolism. This would put together data from government agencies, utilities and industrial sectors, on energy and material flows, to build up a comprehensive framework for strategic and cumulative assessment. This would complement standard in-house environmental management systems for all public bodies and larger organizations.

A green-or-clean technology growth pole would harness regional funding, target skills development, capitalize on resources in the higher-education institutions, finance industrial conversion and technology transfer programmes and develop themed business sites and networks, assisted by green marketing and promotions. Such concepts are at the pilot stage in various locations (Jackson and Roberts 1997). Taken further, such a growth pole would spearhead an environmental business development fund, delivering finance for green investment by accredited businesses. The preferential lending programmes of the Co-op and other banks could be expanded and linked to co-ordinated marketing and distribution schemes.

Larger-scale investment may be essential for environmentally efficient infrastructure such as CHP, light rapid transit, bio-remediation, waste recovery and industrial ecology networks. Such development requires stability in long-term commitments and markets, partnership between sectors, utility regulation that encourages demand-side management and least-cost planning and long-term financial packages which link the interests of purchasers, providers and consumers.

Environmental employment programmes depend on the outcome of the current 'New Deal', but the goal should be to combine environmental and social goals, where appropriate making linkages with the voluntary or third sector of non-monetary services. For direct programmes, such as the upgrading of all buildings in the city to energy best-practice standards, the critical factor would again be the financial package which links lenders, building owners, developers, utilities, welfare and training agencies. First estimates for Greater Manchester indicate a possible 10,000 to 15,000 jobs in a full-scale energy/climate-change programme (Ravetz 1999).

Attempts to apply environmental and ethical policies in public purchasing and contracting suggest the need to form a database of accredited regional manufacturers and suppliers, linked to standard green and/or ethical principles for all public bodies and larger companies. This would promote and underpin other schemes such as integrated chain management and industrial ecology networks in waste, energy and transport: a pilot industrial-ecology scheme is now being established in the Trafford Park development area of Manchester (NCBE 1998).

Such schemes will depend on creative innovation in technical, economic and political systems, responding to opportunities that arise. They will also need a frame of reference, or a set of parameters, for the sustainability of the system as a whole. Such goals offer a profound challenge, and will require actions on all fronts, far beyond the current remit of the RDAs, structural funds, local government and initiatives such as LA21. However, if society is serious about integrating environmental sustainability with economic development there seems to be little alternative.

Conclusion

The chapter has drawn together a number of parameters with which to assess the environmental sustainability of a local economy. It has also considered the degree to which these parameters are reflected in the current situation in Greater Manchester and it has outlined elements of a practical agenda for action. Such a strategy is important because behind the current rhetoric, economic development is still only being greened at the margin in Greater Manchester as in most other cities. However, there is great potential for a proactive environmental management framework based on the city-region as a physical and functional system. Such a framework could have both environmental and economic benefits. Progress depends on operationalizing the city-region as an economic and political unit, and local authorities have a key role to play in this. This raises wider questions on the powers and resources of regional and local government, and their changing relationships with other actors and agencies.

Acknowledgements

I would like to acknowledge the contributions to these ideas from the Sustainable City Region Working Group of the Town and Country Planning Association. The 'sustainable city-region' project has received financial support from the ten local authorities of Greater Manchester Economic and Social Research Council, European Regional Development Fund, DG XII of the European Commission, Manchester Metropolitan University, a private trust, Eversheds, Pieda plc and Ove Arup and Partners.

References

AGMA (Association of Greater Manchester Authorities) (1994) *Strategic Guidance Monitoring Report*: Wigan: AGMA.

Barton, H. and Bruder, N. (1995) *A Guide to Local Environmental Auditing*, London: Earthscan.

Bossel, H. (1996) *20/20 Vision: Explorations of Sustainable Futures*, Centre for Environmental Systems Research, University of Kassel, Germany.

Breheny, M. and Rookwood, R. (1993) 'Planning the sustainable city region' in A. Blowers (ed.) *Planning for a Sustainable Environment*, London: Earthscan.

Carley, M. and Spapens, P. (1997) *Sharing the World: Sustainable Living and Global Equity in the 21st Century*, London: Earthscan.

Clark, M., Gibbs, D., Brime, E. and Law, C. M. (1992) 'The north west', in P. Townroe and R. Martin (eds) *Regional Development in the 1990's*, London: Jessica Kingsley Publishing.

Clayton, A. and Radcliffe, N. (1996) *Sustainability: A Systems Approach*, London: Earthscan.

Co-op Bank (1995) *Ethical Policy*, Manchester: Co-operative Bank.

Douglas, I. and Lawson, N. (1998) 'Translating Local Agenda 21: SME approaches to sustainable development in Manchester', Department of Geography Working Paper, University of Manchester.

DTI/DoE (1994) *The UK Environmental Industry: Succeeding in the Changing Global Market*, London: HMSO.

ECOTEC (1991) *The Implications of Environmental Pressures*, report to Warwickshire County Council, Coventry City Council and the BOC Foundation, Birmingham: ECOTEC.

ECOTE (1996) *Environmental Protection Expenditure by Industry*, report to the DoE, London: HMSO.

Elkington J. (1997) *Cannibals with Forks: The Triple Bottom Line of 21st Century Business*, Oxford: Capstone Publishing.

Entec (1996) *The Application of Environmental Capacity to Land-Use Planning: Seminar Briefing Note*, working paper for the DoE, Warwickshire: Entec.

Environment Agency (1997) *Best Practicable Environmental Option Assessments for Integrated Pollution Control*, technical guidance note. Environmental, E1, London: Stationery Office.

Fothergill, S., Monk, S. and Perry, M. (1987) *Property and Industrial Development*, London: Hutchinson.

Friedman, J. and Weaver, C. (1979) *Territory and Function*, London: Edward Arnold.

Gibbs, D. C. (1994) The Green Local Economy: Integrating Economic and Environmental Development at the Local Level, Manchester: Centre for Local Economic Strategies.

Gibbs, D. C., Longhurst, J. and Braithwaite, C. (1996) 'Moving towards sustainable development: Integrating economic development and environmental management in local authorities', *J. Environmental Planning and Management* 39(3) 317–32.

GMWRA (Greater Manchester Waste Regulation Authority) (1995) *Waste Disposal Management Plan*, Manchester: GMWRA (now Environment Agency).

GONW/GOM (Government Office North West and Government Office Merseyside) (1996) *Regional Planning Guidance for the North West*, London: HMSO.

GMR (Greater Manchester Research, Information and Planning Unit) (1995) *Greater Manchester Facts and Trends*, Oldham: GMR.

Groundwork Foundation (1996) *Small Firms and the Environment*, a Groundwork status report, Birmingham: Groundwork Foundation.

Hitchens, D. (1997) 'Environmental policy and implications for competitiveness in the regions of the EU', *Regional Studies* 31(8), 813–19.

HMIP (Her Majesty's Inspector of Pollution) (1995) *Chemical Release Inventory*, London: HMSO.

Hooper, P. and Gibbs, D. C. (1995) *Profiting from Environmental Protection: A Manchester Business Survey*, report to the Cooperative Bank, Manchester: Dept of Environmental Sciences, Manchester Metropolitan University.

Jackson, T. and Marks, N. (1994) *Measuring Sustainable Economic Welfare – A Pilot Index 1950–1990*, York: Stockholm Environment Institute.

Jackson, T. and Roberts, P. (1997) 'Greening the Fife economy: Ecological modernization as a pathway for local economic development', *J. Environmental Planning and Management* 40(5), 615–30.

Jacobs, M. (1994) *Green Jobs? The Employment Implications of Environmental Policy*, report to the World Wildlife Fund, Godalming: WWF.

James, P. (1994) 'Business environmental performance measurement', *Business Strategy and Environment* 3(2), 59–71.

Lewis, G. J. (1997) 'A Cybernetic View of Environmental Management: the Implications for Business Organizations': *Business Strategy and Environment* 6(5), 264–75.

LGMB (Local Government Management Board) (1994) *Greening the Local Economy*, Luton: LGMB.

March Consulting (1995) 'Multi-client CHP scheme for Trafford Park, Manchester', *Energy Management Journal*, July.

Marvin, S. and Graham, S. (1995) 'Utilities and territorial management in the 1990's' *Town and Country Planning Journal* 63(1).

McNaghten, P., Grove-White, R., Jacobs, M. and Wynne, B. (1995) *Public Perceptions and Sustainability in Lancashire – Indicators, Institutions, Participation*, Lancaster: Lancashire County Council and Lancaster University/CSEC.

NCBE (National Centre for Business and Ecology) (1997) *A Green Competitive Edge for the North West*, working paper, NCBE, Salford University.

NCBE (National Centre for Business and Ecology) (1998) *Trafford Park Industrial Ecology Scheme*, working paper, NCBE, Salford University.

Nijkamp, P., Lasschuit, P. and Soeteman, F. (1992) 'Sustainable development in a regional system, M. Breheny (ed.) *Sustainable Development and Urban Form*, London: Pion.

NWRA/NWBLT (North West Regional Association) (1993) *Regional Economic Strategy for North West England, Regional Transport Strategy for North West England and Environmental Action for North West England*, St Helens: North West Partnership.

ONS (Office of National Statistics) (1996) 'The pilot UK environmental accounts', *Economic Trends*, August, 41–71.

Portugali, J. (1997) 'Self-organizing cities', *Futures* 29(4–5), 358–80.

Pre˙ Consultants (1995) *The Eco-Indicator 1995*, Amersfoort, Netherlands: Pre˙ Consultants.

Ravetz, J. (1996) 'Towards the sustainable city region', *Town and Country Planning* 65(5).

Ravetz, J. *et al.* (1999) *City-Region 2020: Integrated Planning for Long Term Sustainable Development in a Northern Conurbation*, London: Earthscan/Town and Country Planning Association.

Ravetz, J. (2000) 'Integrated assessment for sustainable cities and regions', *Environmental Impact Assessment Review* 20(1).

RCEP (Royal Commission on Environmental Pollution) (1994) *18th Report: Transport and the Environment*, London: HMSO.

Rees, W. and Wackernagel, M. (1995) *Our Ecological Footprint: Reducing Human Impact on the Earth*, British Columbia, Gabriola Island: New Society Publishers.

Richards, L. and Biddick, I. (1994) 'Sustainable economic development and environmental auditing: a local authority perspective', *Journal of Environmental Planning and Management* 37(4).

Roberts, P. (1994) 'Sustainable regional planning', *Regional Studies* 28, 271.

Roberts, P. and Chan, R. (1997) 'A tale of two regions', *International Planning Studies* 2(1), 45–62.

Sustainability North West (1998) *Sustainable Development and the North West Regional Development Agency*, SNW, Manchester University.

Tyteca, D. (1996) 'On the measurement of the environmental performance of firms – a literature review and a productive efficiency perspective' *Journal of Environmental Management* 46, 281.

von Weizsacker, E., Lovins, A. B. and Lovins, L. H. (1997) *Factor Four: Doubling Wealth, Halving Resource Use*, London: Earthscan.

Welford, R. and Gouldson, A. (1993) *Environmental Management and Business Strategy*, London: Pitman Publishing.

Welford, R. (1995) *Environmental Strategy and Sustainable Development*, London: Routledge.

White, R. and Whitney, J. (1992) 'Cities and the environment: An overview', R. Stren, R. White, and J. Whitney (eds) *Sustainable Cities – Urbanization and the Environment in International Perspective*, Colorado: Westview.

Whitelegg, J. (1995) *Freight Transport, Logistics and Sustainable Development*, report to the World Wildlife Fund, Godalming: WWF.

Wolters, T., James, P. and Bowman, M. (1997) 'Stepping stones for integrated chain management in the firm', *Business Strategy and Environment* 6(3), 121–32.

Wood, C. (1996) *Trading in Futures: The Role of Business in Sustainability*, Lincoln: Wildlife Trusts.

13 Integrating environment and economy in a Scottish council

The case of Fife

Tony Jackson

Introduction

Bennett (1997) argues that a 'truly-bounded' local authority, with an administrative system which matches its functional socio-economic space, should make strategic policy-making much easier because:

> if territory, economy, society and environmental impact, as well as place-consciousness are all coterminous, spillover, inter-governmental negotiation, partnership etc, all disappear as issues to be addressed.
>
> (p. 326)

Fife's pursuit of sustainability provides an interesting test of this hypothesis. Successive rounds of local-government reorganization, whilst altering its internal administrative structures, have left Fife's external boundaries unaffected, allowing it to emerge under the latest round as one of Scotland's largest single-tier authorities. This territorial continuity has given Fife a clear spatial identity that overrides wide socio-economic and political divides.

Lacking a dominant city, the three travel-to-work areas (TTWAs) of North-East Fife, Kirkcaldy and Dunfermline cover most of Fife's 340,000 resident population. The 1991 Population Census indicated that net outward commuting amounted to 11 per cent of the 208,000 residential workforce. Although the north-east of Fife remains predominantly rural, the industrial revolution came to the rest of Fife at an early stage, attracted by abundant supplies of water power and coal and easy water-borne transport. The opening up of new coalfields and the establishment of a new town, Glenrothes, led to substantial increases in population in these areas during the early post-war decades.

Since the 1970s, the traditional economic base of the Kirkcaldy and Dunfermline TTWAs has been steadily eroded. By the 1990s, rates of unemployment in Fife exceeded those of any other mainland Scottish region. Following the 1993 review of UK regional assisted areas, the Kirkcaldy and Dunfermline TTWAs were reinstated to full development-area status. This part of Fife also qualified for Objective 2 status (areas of industrial decline) under the present

European Union (EU) Structural Funds programme of regional assistance. Included within the TTWAs are coalmining and dockyards communities which qualify for additional EU RECHAR and RENAVAL assistance.

The additional burden of industrial and environmental dereliction, created by abandoned plant and factories, and the contaminated soil, disrupted drainage, despoiled water courses and disfigured landscape resulting from centuries of deep coal mining and related heavy industry has added to the economic impoverishment of this area. In recent decades, this part of Fife has qualified for a number of major environmental improvement grants, to reinstate derelict and contaminated sites, and to remove abandoned coal bings and other environmental detritus of its traditional activities. The area remains sensitive to continued demands for open-cast coalmining, and the development of petro-chemical activities linked with the North Sea oilfields.

Following its establishment in 1996, Fife Council pursued a set of policies and programmes inherited from its predecessor, Fife Regional Council. This chapter begins its assessment of these efforts by reviewing the concept of ecological modernization and the contribution it may make to the transition toward sustainability. This is followed by an analysis of the policy processes at work which are supplanting the traditional view of policy in which environmental protection was seen as a constraint on economic growth. Subsequent sections examine how changes in management structure, and the adoption of new management techniques and performance measures, have fleshed out the framework of an approach designed to integrate environment and economy. The chapter concludes by examining some of the issues yet to be resolved in Fife's attempt to give the environment a strategic role in its developmental actions.

The environment as a catalyst for economic development

Ecological modernization avows that the pursuit of environmental protection is compatible with the maintenance of economic development (see Hajer 1996). More particularly, it focuses on the potential for new and innovative forms of policy to influence environment–economy relations. However, while its prescriptions for the reform of economic and environmental policies may have potential, it is important to note that ecological modernization does not address some of the main issues that are at the heart of most interpretations of sustainable development. For example, it has been criticized for not giving sufficient emphasis to social issues such as equity and environment–society relations and for not paying enough attention to issues of futurity because it does not address the growth logic inherent in the economic system (see Gouldson and Murphy 1997). Consequently, ecological modernization can perhaps best be seen as one of a wider range of concepts that need to be applied in concert to facilitate a shift towards sustainability.

Ecological modernization discourses are increasingly apparent in real-world policy debates. This can perhaps be seen to best effect in the evolution of environmental action programmes (EAPs) within the EU (see Gouldson and

Murphy 1996). Initially focused on correcting the market failures which allowed economic growth to damage the environment, recent programmes have come to emphasize the role of environmental policy as a facilitator of economic development rather than a burden upon it. A central tenet of the current fifth EAP (CEC 1992) is that future economic development within the European Union requires a shift toward sustainability and therefore the adoption and implementation of new forms of environmental policy.

There are several strands to these propositions. One is that failure to promote environmental protection locks environmental costs into the economic system. These costs have to be met at some point and at some time in ways which will impose long-term economic and social burdens on the communities involved. Another is the claim that high environmental standards provide a competitive edge for some industries, such as tourism or environmental technology, that can exploit the new market opportunities related to the application of stringent environmental standards thereby providing the source for new economic development and employment opportunities. Closely related to this is the claim that environmental regulations can drive innovation, thereby encouraging the development of new technologies and techniques that are both more environmentally benign and economically efficient. The benefits of such innovations, for example in energy efficiency, could be felt throughout the economy and by society at large. The precautionary principle adds a further strand by providing the basis for a more risk-averse strategy in the management of the environment, one which looks on higher environmental standards as desirable in their own right rather than as a cost that has to be met by industry in selling its products.

Although this paradigm has been developed most fully for the manufacturing sector, ecological modernization is of equal relevance to the built and natural environments and to the provision of local public services to their communities. Indeed, it can be argued that the concept of locality is crucial to any effective implementation of ecological modernization, and that bottom-up rather than top-down solutions 'are the distinguishing characteristic of most successful attempts at planning for sustainable development' (Roberts 1996: 5). This brings a spatial and a social dimension to ecological modernization and reinforces the contention of the UK Local Government Management Board (LGMB) that local authorities are uniquely placed to bring together at an appropriate spatial level a set of policies, the means to implement these, and the partnerships to provide the political impetus for sustainability (LGMB 1993).

The evolution of policies for sustainability in Fife

Table 16 is based on a typology of approaches towards sustainability adopted by UK local authorities, suggested by Jackson and Roberts (1997). In 1988, Fife Regional Council adopted an environmental charter, designed to attach corporate responsibility to initiatives dealing with the environment. Under the guidance of an environmental co-ordinator, the final programme under this charter initiative, completed in 1996 just before the regional council along with its three districts

Table 16 Typology of local authorities based on approaches to ecological modernization (Jackson and Roberts 1997)

Environmental fragmentation: the first phase, from which the bulk of UK local authorities have now emerged, characterized by a totally fragmented approach towards environmental policy and programmes. Individual departments may have their own environmental initiatives, but these function within a management system lacking any coherent means of incorporating environmental aspects into the overall decision-making process.

Embryonic environmental greening: the second phase, characterized by an acknowledgement of the need to have a corporate environmental stance. Although environmental measures may still consist largely of unco-ordinated individual projects, these local authorities have undertaken a review of their activities in the light of their commitment to pursuing a more environmentally conscious profile. This phase includes the promotion of corporate initiatives, such as recycling targets and natural conservation strategies, which are essentially retrofitted onto existing corporate policies, and the appointment of an environmental co-ordinator. Most UK local authorities have reached at least this point of development.

Corporate environmental greening: the third phase, which includes an increasing proportion of UK local authorities, characterized by a clear corporate strategy on environmental issues, involving the integration of environmental policies with other corporate objectives. Markers for this phase are the appointment of environmental co-ordinating committees to implement sustainability programmes and the introduction of management techniques designed to formulate and implement corporate environmental targets. Local authorities in this phase will be producing corporate reports on their environmental activities, which compare achievements against corporate commitments to the environment

Strategic environmental greening: the final phase, which the most committed UK local authorities are currently exploring, characterized by full-scale re-engineering designed to structure both the corporate objectives and their delivery around the concept of sustainable development. This involves a re-examination of the aims of the local authority, adoption of management structures favouring the development of strategic approaches to sustainability, and the use of performance measures to develop and evaluate spending programmes on an environmental basis. One of the markers for the start of this phase is the development of indicators of sustainability. Another is the restructuring of the decision-making process, with a view to making it far more accessible to local communities. This approach is akin to the implementation of an ecological modernization strategy

were merged into a single-tier authority, Fife Council, addressed sixty-six environmental projects over a 2-year period. Although many of these were one-off measures operating within an unchanged policy framework, the programme attracted various environmental awards, including the title of 'Scottish Green Local Authority of the Year', and placed the region firmly in an embryonic greening phase.

Under the former regional council the environmental programme, rather than being incorporated into other council policies, ran parallel with them. The 1994 structure plan (FRC 1994) still shows little change from the traditional paradigm.

Although it devotes a separate chapter to the environment, this was essentially 'retrofitted' to an independently worked economic and social strategy for development of the region. An effort is made to link environmental initiatives to socio-economic objectives, but no attempt is made to undertake a strategic environmental assessment of the impact of various proposals. At the time, the council lacked the means to contemplate this, even if it possessed the political desire.

The legacy the regional council left for its successor consisted of a set of initiatives designed to move Fife Council into the next phase, including adoption of a clear corporate strategy on environmental issues, revision of the charter to incorporate sustainability, the introduction of a state-of-the-environment report linked to an environmental database for Fife, and creation of a Fife business forum for the environment. At the same time, as the following sections outline, steps were being taken to modify management structures and introduce management techniques to provide the means to realize these ambitions.

At this point, a new driving force began to provide additional impetus to the policy process. By 1994 Local Agenda 21 (LA21) had been enthusiastically endorsed by UK local government. Based on the concepts championed by the Brundtland Commission (WCED 1987) and the World Bank (IBRD 1992) and adopted by the 1992 UN Conference on Environment and Development, Agenda 21 offers a detailed programme of actions favouring measures of development which comply with intra- and inter-generational equity and environmental considerations. Chapter 28 of Agenda 21 argues that the participation and co-operation of local authorities will be a determining factor in meeting the requirements of sustainable development, because so many of the problems and solutions involved have their roots in local activities. Furthermore, because local authorities provide the level of government closest to the people, they can help to educate the public in sustainability, mobilize public awareness and opinion on its importance and respond to public needs in its promotion.

The LGMB identifies six essential elements for implementing LA21:

- managing and improving the local authority's own environmental performance;
- integrating sustainable development aims in the local authority's policies and activities;
- raising awareness of and promoting education about sustainable development;
- consulting and involving the public;
- creating partnerships;
- measuring, monitoring and reporting on progress towards sustainability (LGMB 1994).

The sustainability policy (FRC 1996) which emerged shortly after the new council was formed is a re-working of the original regional council environmental charter to incorporate sustainability objectives. Perhaps understandably, neither the policy document nor the first action plan derived from it attempt to address the strategic implications for current decision-making processes and service delivery arrangements (Bruder 1997) of applying an approach to council actions which matches the UN sustainability agenda. Instead, the policy document is intended to serve a fourfold functional purpose:

- to be a public statement of the council's commitment to sustainability;
- to offer its council officers guidance on how to incorporate sustainability in council policies and programmes;
- to act as an awareness-raising document;
- to provide an input into the preparation of Fife's LA21.

The document is based on a set of sustainability principles (see Table 17) and a set of procedures for implementing these. Both are based on guidance derived from nationally agreed documents. The procedures reproduce the six steps laid down by the Local Government Management Board for implementing LA21. Guidance on how to construct specific objectives meeting these principles is absent. Instead, the emphasis is on corporate co-ordination to deliver over time the detailed measures originated through individual service action programmes. The first action plan for delivering this policy (FRC 1997) consists of a list of initiatives grouped under the LGMB's six steps for implementing LA21.

A key feature is the emphasis given to getting local interest groups involved in helping to formulate action programmes for sustainability. In Fife, a roundtable has been established to bring these together into a partnership. Most of the partners operate on a Fife-wide basis, reinforcing Bennett's observation about true-bounding. The first roundtable report (SFR 1996) reviews the council's own

Table 17 Fife Council Sustainability Policy statement (FRC 1996)

Fife Council is committed to the concept of sustainability. We aim to improve the quality of life and quality of the environment in Fife by providing for individual and community needs, whilst ensuring availability of natural resources now and in the future. The following principles of sustainability will be adopted in all of the council's activities, including policy formulation and the delivery of services, now and in the future:

1 Minimize the use of finite resources
2 Promote access to worthwhile and productive jobs
3 Promote social equity
4 Protect and improve quality of life by meeting people's needs for amenities and services locally
5 Ensure strong safe and thriving communities
6 Conserve and enhance biodiversity
7 Protect and enhance the visual landscape and townscape
8 Raise environmental awareness and education
9 Improve public consultation and participation processes
10 Promote high standards of health

sustainability policy, sets out the management framework through which the partnership will operate to influence this, and provides a brief statement on the LA21 actions already being taken by participants.

The statement on current actions reveals a wide disparity in perception amongst the partners both about the aims of sustainability and about the role of participants in achieving these. In addition to the local authority, most of the other public-sector institutions involved have already adopted their own sustainability policies. In the case of Scottish Natural Heritage and the Scottish Environmental Protection Agency, these are well integrated into their corporate strategies. Voluntary-sector organizations offer a fragmented approach towards sustainability. The statement from the higher-education sector refers purely to the service impacts of its environmental research and teaching, ignoring the direct impacts on local resources its own rapid growth is producing. Manufacturing and commerce see their interests represented by the Green Business Fife initiative started under Fife Regional Council, while farming and fishing adopt roles as guardians of the environment. Fife Enterprise, the principal agency charged with local economic development, regards sustainable development as an important secondary but subsidiary objective to its main focus on job creation.

Such a diversity of views is understandable at the outset of a policy initiative. However, if the roundtable is to play a constructive role in formulating and promoting LA21 policies for Fife, its participants will need to form a common understanding of the meaning of sustainability and of its implications for their own activities. This will require some careful guidance on the part of the local authority.

Managing sustainability: Changes in the structure of local authority decision-making

The adoption by Scottish councils of sustainability policies has coincided both with major changes in the structure of local government and, more fundamentally, with a prolonged period of critical re-evaluation of their role. Clark and Stewart (1994) postulate four possible models of future community governance:

- the unitary authority reverses recent trends and reinstates the local council as chief provider of local public services based on local choice;
- the residual authority recognizes recent trends by removing further functions from the elected council, leaving it with the bare minimum of responsibilities and placing community governance in the hands largely of a mixture of delegated central-government agencies and market forces;
- the advocacy authority, despite the continued removal of further functions, retains the ability to make representations on behalf of the local community, replacing local choice with local voice;
- the strategic authority switches from provider of local public services to the management of the local community's interests in these, with consensus

seeking between non-elected local delivery agencies and local communities becoming the focus of community governance.

As Clark and Stewart acknowledge, these models are not necessarily mutually exclusive. At the same time as losing important responsibilities such as water and sewerage to government agencies, Fife Council has acquired for the first time full unitary control over its remaining functions, subject to increasingly stringent compulsory competitive tendering regulations. Overall, Fife Council's corporate response to local-government restructuring has strengthened its role as a strategic authority. This is particularly evident with respect to its management of sustainability policies.

Figure 7 illustrates the radically different management structure instituted by the new authority in 1995. This is designed to improve corporate decision-making and to expand the scope for decentralized delivery of services. Three key areas of strategy are identified, namely social, environmental and developmental and competitive and technical.

The appointment of corporate managers for these three areas who report directly to the Chief Executive allows professional services to be co-ordinated along lines which enable coherent strategies such as LA21 to be introduced.

In addition to these corporate managers, the authority has created a Citizenship Commission at full council level charged with taking forward and developing decentralization, consultation and community and citizen involvement in council affairs. Underpinning this is a policy of strategic decentralization through the creation of a number of local area co-ordinators in charge of multi-functional area offices delivering local community services and empowered to work across departmental lines. Each of these areas also has an area committee

Figure 7 Strategic management organization for Fife Council (Jackson and Roberts 1997)

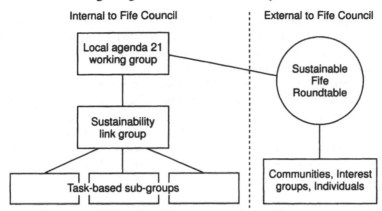

Figure 8 A framework for Local Agenda 21 in Fife (SFR 1996)

composed of elected councillors for the area, with devolved responsibilities for determining local decisions.

Figure 8 illustrates the arrangements for co-ordination of sustainability policy, which supplement these internal management reforms. Horizontal integration of sustainability actions across departmental lines is promoted by an LA21 working group of members and officers with links to departmental service providers and to the roundtable partnership. The internal links are channelled through a council officer committee termed the Sustainability Working Group, to which specific task-based subgroups report. Service department action programmes are scrutinized by the sustainability policy subgroup checking for conformity with the corporate commitment to sustainability.

These management reforms illustrate a revival of interest in the systems approach towards management, which is better able to meet the holistic demands of sustainability than the hierarchical structures common to most local authorities. Dryzek (1987) argues that hierarchical management systems are intrinsically poor at coping with the complexity and speed of change of human ecosystems. A basic tenet of the systems approach is that the outcomes of any complex system are often counter-intuitive: the more complex it becomes, the more a system's overall behaviour depends on interaction between its different elements rather than on the initial elements themselves. It follows that in delivering desired outcomes such as sustainability, not only must the individual elements in the system be managed in a sustainable fashion, so must the system itself (Brugman 1992).

The Expert Group on the Urban Environment (CEC 1996) identifies the following characteristics which make hierarchical structures ill-suited to the delivery of systematic policies of sustainability:

- sectoral specialism of individuals, frustrating the ability of the individual to gain an overview of the operation of the system as a whole;
- sectoral specialism of organizations, which creates the same problems for individual departments;

- inappropriate quantification of performance, measuring the success of individuals and departments by factors unrelated or negatively related to sustainability (e.g. kilometres of road maintained): the 'tonnage ideology' measures to which Simonis (1993) refers;
- undue emphasis on the marketability of services, reinforcing financial targets and discounting environmental externalities related to these (e.g. promotion of cost centres which fail to include all ancillary impacts on the community of changes in modes of delivery, so that 'purchasers' define their service requirements as tightly as possible and 'providers' deliver these as cost-effectively as possible, both ignoring the non-attributable environmental impacts).

For a systematic approach to sustainability, a management system must be capable of incorporating the following principles (CEC 1996: 55):

- vertical and horizontal integration of organizations, policies, plans and programmes;
- co-operation and recognition of mutual dependence between all agents within the system, based on a proactive approach to consensus building and networking;
- promotion of homeostasis, so that the management of dynamic change operates within a flexible but broadly stable system;
- application of subsidiarity, allowing decision-making to function at the lowest level consistent with achievement of goals;
- acknowledgement of synergy, recognizing that the whole is greater than the sum of the parts in the creation and implementation of strategic spatial plans.

Delivering sustainability

The two previous sections outlined the progress made in Fife in implementing three of the six steps of the Local Government Management Board's Sustainability Action Programme (LGMB 1994): integrating sustainable development aims into council policies and activities; working together through partnerships; and awareness-raising. This section considers the remaining three steps: the management of improved environmental performance within the council itself; measuring, monitoring and reporting on progress towards sustainability; and consulting and working with the general public. If the first three steps could be considered as issues of design and development, then the remaining three are matters of delivery and performance.

Environmental management systems

The key to delivery of a local authority's environmental objectives performance lies in effective application of an environmental management system. EU policy towards such systems has evolved from proposed obligatory directives to voluntary schemes. The initial 1993 Eco-Management and Audit Scheme (EMAS) regulation was restricted to corporations performing industrial activities on registered sites. It received a more enthusiastic reception in the UK, where corporate environmental standards are generally management-driven, than in some continental member states such as Germany, where standards are technology-driven (Franke and Waetzold 1996).

The existence of well-established eco-auditing procedures in the UK quickly encouraged the development of an EMAS for local government (BIE 1994). This uses a standard management loop (establish policy, review current position, set objectives, implement targeted programme, monitor performance and adapt policy, objectives and/or programme as required) to integrate environmental strategy into corporate management. There are two innovative features which distinguish this approach from other standards. The first, which applies both to the original scheme and the local-government version, is a validation arrangement in which an independent verifier inspects the undertaking to validate compliance with the regulation. The validation process is based on the undertaking's publicly available environmental statement. Both the statement and the independent validation process go beyond the BS 7750 and ISO 14001 standards, although the processes up to this stage are complementary.

The second applies purely to the local-authority version. This extends the scheme from single-site registration and validation of the direct effects of an undertaking's production activities, to multi-site registration and validation of all an authority's environmental impacts, including the service effects of its policies as well as the direct environmental impacts of its operations. These requirements turn the standard into a much more ambitious environmental management system. For their own internal purposes, local authorities may voluntarily adopt EMAS on a piecemeal basis within individual service units, for functional tasks (purchasing, transport, etc.), or for specific types of inputs (energy, water, paper, etc.). However, registration of such an exercise renders each step in the process compulsory and also requires the authority to have a corporate overview and co-ordination system for environmental management. In addition, the whole authority must commit itself to corporate registration and validation at some future time. The local-authority scheme is broader not only in scope, but also in methodology. While under the original scheme, one would normally expect direct effects to have adverse environmental impacts requiring to be minimized, under this much wider approach the scheme has to accommodate the likelihood that some of the service effects of an authority will be environmentally beneficial.

In 1995 Fife became a pilot authority for testing the local authority EMAS as one of four European authorities participating in an EU Partnership in EMAS (PIE) project. The council focused on four functional areas of management, rather than specific material inputs. Supplies and transport allow the effectiveness of the scheme to be examined in areas where direct effects predominate, and economic development and planning where service effects create the main impact. The use of EMAS has allowed the authority to switch its emphasis away from a project-driven environmental agenda towards one which focuses on the management processes required to implement sustainability.

One result of Fife's experience with EMAS is to confirm that such approaches operate more effectively under the systematic corporate approach to management adopted by Fife Council than the hierarchical management structures of its predecessor. The testing of EMAS has confirmed the difficulty of applying corporate-wide environmental strategies within hierarchical structures. Senior management encounters problems in integrating services across departmental lines, because of the line management systems common to professional services. Hill and Smith (1994) argue that a traditional local-government structure:

> may give rise to a fundamental contradiction of values and objectives between 'professionals' who see their prime duty to aspects of the 'environment' and others whose concern is the effectiveness of the whole service. Although there are some advantages to this [traditional] model, it has several disadvantages that are likely to militate against achieving sustainability and its implementation through environmental auditing and management.
>
> (p. 91).

Currently Fife Council is completing its reviews of the piloted functions and preparing its own version of an environmental management manual.

Sustainability Indicators

Although corporate management systems can ensure that sustainable policies are implemented, they cannot by themselves generate such policies. Nor can they measure the effect of these policies on the external environment. Chapter 40 of Agenda 21 notes that indicators of sustainable development need to be developed to provide solid bases for decision-making and to contribute to the self-regulating sustainability of integrated environmental and socio-economic development systems. One of the main LA21 research initiatives promoted by the LGMB has been to develop suitable procedures to this end (Touche Ross 1994 and 1995). The UK sustainable development strategy (DoE 1994) draws a distinction between indicators of sustainability and indicators of environmental quality. The strategy argues that in addition to assessing the state of the environment, sustainability indicators must take account of economic linkages, the quality of life and future welfare aspects. In a pilot project involving Fife (Touche Ross 1994 and 1995; FRC 1995a and 1995b), the LGMB tested the effectiveness of indicators derived from thirteen previously agreed sustainability themes based on this strategy, which are summarized in Table 18.

The themes were chosen to incorporate the UNEP working definition of sustainable development, which seeks to improve the quality of life within the carrying capacity of supporting ecosystems. Thus the first three themes deal with carrying capacity, while quality of life is taken to mean meeting local needs for basic necessities, health, access to facilities, work, freedom from crime and

Table 18 Requirements for a sustainable community: Themes for Sustainability
Indicators Research Project (Touche Ross 1994)

Theme	Meaning
Resources and waste	Efficient use of resources and waste minimization by closing cycles
Pollution	Limitation of pollution to levels within the carrying capacity of natural systems
Biodiversity	Valuing and protecting the diversity of nature
Localness	Meeting needs locally where possible
Access to basic needs	Offering everyone access to good food, water, shelter and fuel at reasonable cost
Work	Providing everyone with the opportunity to undertake satisfying work in a diverse economy; recognizing the value of unpaid work; offering fair and fairly distributed payments for work
Health	Protecting people's good health by creating safe, clean, pleasant environments and health services which emphasize prevention as well as proper care for the sick
Access to facilities	Ensuring access to facilities, services, goods and other people is not achieved at the expense of the environment or limited to those with cars
Crime	Offering people opportunity to live without fear of personal violence from crime or persecution because of their personal beliefs, race, gender or sexuality
Access to skills and knowledge	Access to skills, knowledge and information to enable everyone to play a full part in society
Empowerment	Empowering all sections of the community to participate in decision-making
Culture and recreation	Making opportunities for culture, leisure and recreation readily available to all
Aesthetics	Combining meaning and beauty with utility in places, spaces and objects; providing settlements that are human in scale and form; valuing and protecting diversity and local distinctiveness

persecution, involvement in the community, etc. The six pilot authorities had the task of testing indicators derived from these themes, to see whether it was possible to identify meaningful measures which were both practicable and acceptable locally. In some cases, there was a clear trade-off between public acceptance and scientific rigour. Of the 101 possible indicators suggested by the consultants, ninety-five were chosen by piloting authorities, but only seven were chosen by seven or more of the ten piloting authorities. Many were adapted and forty-five additional ones were developed.

As well as testing indicators on a region-wide basis, Fife chose to examine community involvement in sustainable development in three specific locations: the East Neuk fishing communities, Glenrothes New Town and Benarty ex-coalmining communities. The project was seen as providing the opportunity to measure the quality of life and environmental conditions in the region. Public

Table 19 Summary of trends in chosen Fife sustainability indicators (FRC 1995a)

Away from sustainability	*No clear trend*	*Towards sustainability*
Homelessness	Homelessness	Life expectancy
Long-term unemployment	Long-term unemployment	Infant mortality
Poverty	Poverty	Nursery education
Land quality	Land quality	Pedestrian and pedal-
Biodiversity	Biodiversity	cyclist safety
Quality of surface and underground water	Quality of surface and underground water	
Pleasant urban environment	Pleasant urban environment	
Food supply: agriculture	Food supply: agriculture	
Food supply: fisheries	Food supply: fisheries	
Energy	Energy	

consultation identified four of the thirteen themes as being of greatest significance: access to basic needs, health, crime and pollution in order of importance (crime was not included as an indicator by Strathclyde, one of the four shadow authorities which also undertook the exercise).

Fife's final report on this project includes twenty indicators for sustainability which as Table 19 indicates can be sorted into three categories: those which involve a departure from sustainability, those which indicate a movement towards sustainability, and those with no clear trend. Findings are set out using the Sustainable Seattle layout and are illustrated where possible with graphic trends. For the three communities specially chosen for involvement, the following questions were posed to enable local actions towards sustainability to be formulated:

- are the basic needs of local people being met?
- does the lifestyle of local people compromise the ability of people in other places to meet their needs?
- are local people being encouraged to improve their quality of life?
- is the environment being cared for to ensure that future generations are able to meet their own needs?

One of the by-products of this exercise was the community involvement achieved, so meeting another of the steps suggested by the LGMB for implementing LA21. Lessons learnt from the exercise included the difficulty in communicating to people what was meant by the LA21 version of sustainable development. Against this, the inclusion of quality-of-life indicators with ecological ones gave scope to bring into partnership community groups which hitherto were considered to have a purely social function.

Continued development of the indicators is being linked to the incorporation of LA21 into council structures through the authority's sustainability policy.

Efforts are being made to identify effective linkages between the indicators and other programmes using datasets available at local level. This will allow correlations between variables to be explored, making use of Geographical Information System packages to map the findings. This will allow a number of the indicators to be tested for validity against other measures which have already been evaluated. It will also permit a spatial analysis of the indicators to be undertaken, examining the sub-regional spread and location of indicators defined at regional level. A further purpose of this research is to evaluate relationships between socio-economic and environmental factors, and to carry out work on rural deprivation.

Clearly, in the absence of suitable indicators, effective development and implementation of sustainability measures will continue to pose difficulties, even if Fife adopts both a clear sustainability policy process and a corporate environmental management system. Equally, as Brugman (1997) argues, even well-developed indicators can fail to meet their objectives if they are unrelated to a methodical planning process. Brugman suggests that while at local level indicators are vital for performance measurement, they are a sub-optimal tool for technical assessment and even for public education. Attempts to achieve multiple objectives from such indicators compromise their performance measurement function. Brugman's concerns about the use of sustainability indicators are related to:

> the common notion that the measurement of sustainability can be achieved through a public participation process. (p. 62).

They are based on the complexity involved in characterizing sustainability, which necessitates the following indicator requirements:

- integrating indicators, linking economic, social and environmental phenomena;
- trend indicators, linking targets and thresholds;
- predictive indicators, dependent on forecasting models; or alternatively, conditional indicators, applying alternative scenarios;
- distributional indicators, measuring equity effects on an intra- and inter-generational basis through the use of disaggregated data;
- condition–stress–response indicators, offering simple causal models for local conditions (Maclaren 1996: 186).

Brugman suggests that the Sustainable Seattle exercise failed to address the 'bellweather test of sustainability', which requires a definition of the condition any indicator is intended to measure, and that simplicity and participation triumphed over complexity and depth of understanding. Rather than emphasize the strategic or educational or aspects of a sustainability-indicator exercise, he recommends that such tools be confined to performance measures. Adopting this approach, technical and comprehensive assessment of environmental impacts would be prepared through a modified state of the environment review, using the tools of strategic environmental assessment and environmental impact analysis, and educational elements would be given separate attention. In an interesting review of the experience of several US municipalities, he concludes that:

> The most results-oriented indicators projects are those that use indicators to hold institutions accountable to their plans and to evaluate whether their actions are having the desired effects.
>
> (Brugman 1997: 71).

At this stage, the extent to which Fife is employing sustainability indicators as an integral management tool for a performance-driven programme of sustainability measures is yet to be established. Amongst the forty-five initiatives in the first sustainability action programme are the development of environmental quality indicators to assist in the assessment of a state-of-the-environment review for Fife, and the development of a set of technical indicators to measure the impact of the new structure plan on sustainability.

Conclusions

In the space available, it is impossible to do justice to the range of activities and initiatives that make up Fife's commitment to sustainable development. The focus of this chapter has been on Fife's progress towards the final phase of environmental greening and the effective design, development, delivery and performance measurement of an integrated programme based on sustainability.

Although essential, the policy processes facilitating the adoption of sustainability are a relatively uncomplicated element of the whole exercise. The various partnership arrangements, community education and awareness initiatives require little of local authorities that they are not already well-versed in doing. Much more daunting are the changes in corporate structures and corporate performance measures needed to deliver the objectives of sustainability. The pursuit of sustainability provides an excellent test of the ability of any organization to think and operate on a corporate level. It requires a local authority to view its activities as a coherent whole, using systematic management structures and operational tools to adopt sustainable practices.

Echoing Brugman, the key to the process lies in getting the management structures right. This will allow the use of appropriate performance indicators to measure the effectiveness of corporate efforts at sustainability, providing the environmental management system with the feedback to incorporate necessary amendments to programme, policy or objective. Though Fife has maintained the political momentum necessary to introduce sustainability as a corporate objective, it remains to be seen whether this can be translated into corporate management practices and performance indicators.

The existence of an active LA21 forum in Fife has an added virtue in this respect. As Bruder (1997) observes, the technical requirements of environmental management systems can detract both from the participatory aspects of sustainability and the capacity of local authorities to implement sustainability in areas difficult to assess and quantify. This is not true of LA21. Here the emphasis is on consensus building, empowerment and strategic management. Research on the way in which local partnership networks help establish a critical mass of support for commonly accepted policy goals (Bennett and Krebs 1994) suggests that LA21 fora can play an important part in committing development agencies within the partnership to a strategic and dynamic approach to sustainability. Together, the combination of the authority's technical capacity and expertise, and the participatory elements of LA21 offers promise that the new paradigm has a robust future.

A further factor which has assisted the pursuit of sustainability in Fife is the presence of Fife-wide community-interest groups and partners. In other parts of Scotland, local government is just beginning to emerge from the logistical problems of reorganization, and the need to establish new linkages and networks. This has served as a brake on progress towards sustainability. New city authorities have lost much of their strategic hinterland, and some new rural authorities function predominantly as dormitories for commuters. In both cases, effective delivery of sustainable policies will involve complicated inter-authority strategies, plans and programmes. In Scotland, hitherto familiar with a two-tier local-government system, in which structural planning was a top-tier function, this will involve a new learning curve. Fife's experiment offers support for Bennett's hypothesis about the value of truly bounded authorities for strategic policy-making.

References

Bennett, R. J. (1997) 'Administrative systems and economic spaces', *Regional Studies* 31(3), 323–36.

Bennett, R. J. and Krebs, G. (1994) 'Local economic development partnerships: An analysis of policy networks in EC-LEDA local employment development strategies', *Regional Studies* 28(2), 119–40.

BIE (Business in the Environment) (1994) *A guide to the Eco-Management and Audit Scheme for UK Local Government*, London: Business in the Environment.

Bruder, N. (1997) 'Lessons from environmental auditing for the development of local environmental policy', in S. M. Farthing (ed.) *Evaluation of Local Environmental Policy*, Aldershot: Avebury.

Brugman, J. (1992) *Managing Human Ecosystems: Principles for Ecological Municipal Management*, Toronto: International Council for Local Environmental Initiatives.

Brugman, J. (1997) 'Is there a method in our measurement? The use of indicators in local sustainable development planning', *Local Environment*, 2(1), 59–72.

Clark, M. and Stewart, J. (1994) 'The local authority and the new community governance', *Regional Studies* 28(2), 201–7.

CEC (Commission of the European Communities) (1992) 'Towards sustainability, a European Community programme of policy and action in relation to the environment and sustainable development', COM(92)23, Brussels: Commission of the European Communities.

CEC (Commission of the European Communities) (1996) European Sustainable Cities – Final Report of the Expert Group on the Urban Environment, Brussels: Commission of the European Communities.

DoE (Department of the Environment) (1994) Sustainable Development: The UK Strategy, London: HMSO.

Dryzek, J. (1987) Rational Ecology: Environment and Political Economy, Oxford: Basil Blackwell.

Franke, J. F. and Waetzold, F. (1996) 'Voluntary initiatives and public intervention – the regulation of eco-auditing', in F. Leveque (ed.) Environmental Policy in Europe: Industry, Competition and the Policy Process, Cheltenham: Edward Elgar.

FRC (Fife Regional Council) (1994) The Fife Structure Plan, Glenrothes: Fife Regional Council.

FRC (Fife Regional Council) (1995a). Sustainability Indicators Project: Summary and Recommendations, Glenrothes: Fife Regional Council.

FRC (Fife Regional Council) (1995b) Sustainability Indicators for Fife: Measuring the Quality of Life and the Quality of the Environment in Fife, Glenrothes: Fife Regional Council.

FRC (Fife Regional Council) (1996) Sustainability Policy, Glenrothes: Fife Regional Council.

FRC (Fife Regional Council) (1997) First Sustainability Action Programme, Glenrothes: Fife Regional Council.

Gouldson, A. and Murphy, J. (1996) 'Ecological modernisation and the European Union', Geoforum 22, 1.

Gouldson, A. and Murphy, J. (1997) 'Ecological modernisation: Restructuring industrial economies', in M. Jacobs (ed.) Greening the Millennium? The New Politics of the Environment, Oxford: Blackwell.

Hajer, M. A. (1996) 'Ecological modernisation as cultural politics', in S. Lash, B. Szerszinski and B. Wynne (eds) Risk, Environment and Modernity: Towards a New Ecology, London: Sage.

Hill, D. and Smith, T. (1994) 'Environmental management and audit', in J. Agyeman and B. Evans (eds) Local Environmental Policies and Strategies, London: Longman.

IBRD (1992) Development and the Environment: World Development Report 1992, Oxford: Oxford University Press.

Jackson, T. and Roberts, P. (1997) 'Greening the Fife economy: Ecological modernisation as a pathway for local economic development', Journal of Environmental Policy and Management 40(5), 617–33.

LGMB (Local Government Management Board) (1993) A Framework for Sustainability, Luton: Local Government Management Board.

LGMB (Local Government Management Board) (1994) Local Agenda 21: Principles and Process, A Step-by-Step Guide, Luton: Local Government Management Board.

Maclaren, V. W. (1996) 'Urban sustainability reporting', Journal of the American Planning Association 62(2), 184–95.

Roberts, P. (1996) 'Ecological modernisation approaches to regional and urban planning and development', paper given at the International Conference on Environment, Planning and Land Use, Planning and Environment Research Group, University of Keele, April.

SFR (Sustainable Fife Roundtable) (1996) *Local Agenda 21: Working Towards a Sustainable Fife*, Glenrothes: Sustainable Fife Roundtable.

Simonis, U. E. (1993) 'Industrial restructuring: Does it have to be jobs versus trees?', *Work in Progress of the United Nations University* 14(2), 6.

Touche Ross (1994) *Sustainability Indicators Research Project: Report of Phase 1*, Luton: Local Government Management Board.

Touche Ross (1995) *Sustainability Indicators Research Project: Consultants' Report of the Pilot Phase*, Luton: Local Government Management Board.

WCED (1987) *Our Common Future*, Oxford: Oxford University Press.

14 Consensus-building for environmental sustainability
The case of Lancashire

Joe Doak and Angus Martin

Introduction

One of the key features of the drive for sustainable forms of development is the emphasis placed on holistic analysis and integrated action to tackle environmental, economic and social problems. This holism needs to be managed through a process which expands the range of people and interests who are responsible for determining how these problems are defined and how they should be tackled. Figure 9 provides a definitional framework for sustainability which highlights these two aspects. This chapter concentrates on the way in which the principle of participation in sustainable development has been applied in one particular case. It also raises lessons and issues which have general applicability and which ask difficult questions of attempts to integrate economic and environmental policies and practices through participatory decision-making processes.

One particular mechanism for achieving effective public involvement that has come to prominence in recent years has been that of 'consensus-building'. Consensus-building identifies areas of mutual gain and attempts to construct 'win-win' outcomes from the decision-making process. It builds on earlier attempts at community-based planning but attempts to expand the range of interests involved in order to address and resolve conflicts and develop an integrated approach to policy and action across a range of policy fields. However, that process takes place in a political and economic context which has been structured by significant inequalities in power between individuals, social groups and organizations. The assumption that underpins this coming-together of unequal and conflicting parties is that there is one objective which everyone should be able to agree on: the requirement for all forms of economic and social development to be environmentally sustainable.

This chapter examines the experience of the Lancashire Environmental Forum and the production of the Lancashire Environmental Action Programme (LEAP) as a consensus-building exercise. It begins by outlining a normative model of what consensus-building should involve; describes the application of these ideas in the Lancashire example; and then critically evaluates that experience both against the 'ideal' approach and in relation to more analytical ideas developed as part of a wider political economy of policy-making and implementation.

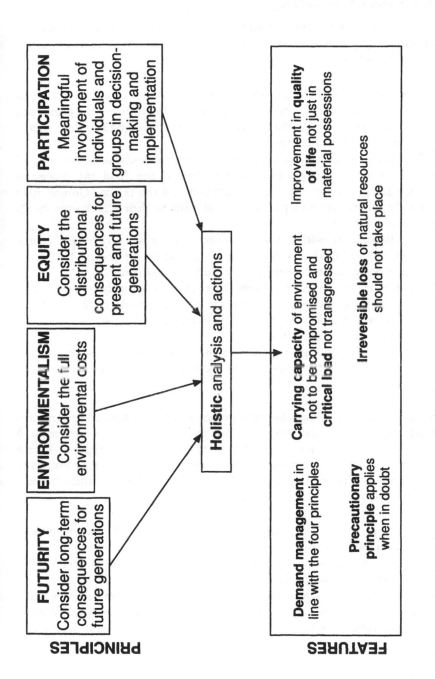

Figure 9 The main principles and features of sustainable development (based on Shorten 1993 and FoE 1994)

What is consensus-building?

In simple terms Acland (1993) describes consensus-building as a process which involves getting people together to talk about an area of mutual concern. He identifies five stages:

- Assessment and preparation – the process begins with a thorough assessment of the current position. The existing processes, the issues and the people and parties to be involved in the process should be considered and these should be built into the participation strategy.
- Initiation – the assessment and preparation stage should give an indication of how to initiate the process. How to launch and fund the process is important, as is the venue for negotiation and the timing. Ground rules for the process and how participants will relate to their constituents should also be considered.
- Negotiation, exploration and exchange – there is a need for neutral facilitators and mediators. Shared goals should be developed. The means of generating and evaluating options must be considered.
- Decision-making – once options have been thoroughly explored it is necessary for commitments to be made. Options should be 'reality tested' in order to determine who pays and implements them. The chosen options must be agreed in some way.
- Maintenance – once people are committed, the process must continue to function until at least some of the proposed solutions are known to be working.

The essence of the consensus-building process is that a negotiated agreement is not an end in itself. The process is just as important as the agreement it produces, especially if the agreement is to achieve the objectives of the participants, gain the commitment of implementers and provide for constructive relationships and future negotiation. The benefits of the consensus-building approach include:

- it allows participants to reach a common understanding of the rather difficult and politically contested concept of sustainable development;
- it aids implementation by providing a better sense of ownership than traditional top-down policy impositions; and
- it mobilizes wider support and concentrates on long-term solutions which reduces the tendency toward the narrowly focused short-termism of party-political policy-making, government implementation programmes or market processes.

Others have been less optimistic about the contribution of consensus-building techniques and processes (Hardy 1990 and Marshall 1994). They argue that it:

- smothers the inherent conflicts through the provision of bland, diluted compromise positions;
- involves covering-up the inequalities in power between participants;
- lacks the institutional framework (in the UK, at least) to achieve its objectives; and
- lacks continuity; consensus being a temporary and fragile social construct.

In order to evaluate this critique of consensus-building, it is useful to explore a recent attempt to progress environmental sustainability using this approach.

Consensus-building: The practice

The approach used in Lancashire, in which a forum has been used to formulate and implement an environmental action programme, has received international recognition (ICLEI 1995). The Lancashire Environmental Forum was established by Lancashire County Council in December 1989. It now comprises over ninety organizations drawn from national-government departments and agencies, industry and unions, local government, interest groups and academic establishments (see Figure 10). Membership is voluntary. Forum members have provided the basis for data collection, acted as the main decision-making mechanism and are now the key instrument through which proposals are being implemented.

The new environmental agenda for Lancashire has been progressed in four stages (see Figure 11):

- stage 1 – the production of a state-of-the-environment report for the county, 'Lancashire: A Green Audit', in 1991;
- stage 2 – the production of Lancashire Environmental Action Programme (LEAP) in 1993; this is a consensus-orientated Local Agenda 21;
- stage 3 – implementation of the 203 proposals in LEAP over periods of between 5 and 15 years;

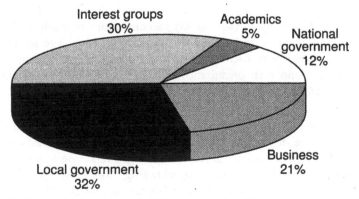

Figure 10 Lancashire Environmental Forum membership

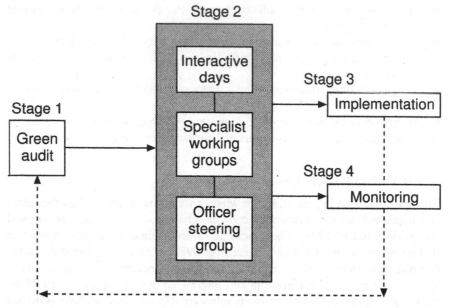

Figure 11 The LEAP process

- stage 4 – monitoring and review of both the green audit and LEAP.

The LEAP was produced over 3 years. The main mechanisms used were:

- interactive days which, due to the large numbers involved, were only held about twice a year. These are formal meetings, held at the county council and chaired by a leading county councillor;
- Specialist Working Group (SWG) meetings held outside the county council and consisting of ten to fifteen individuals functioning as an informal workshop. The SWGs used brain-storming methods to generate ideas and provoke discussion;
- the Officer Steering Group (OSG) meeting more regularly and, like the SWGs, operating on an informal and participative basis. The agenda for the OSG meetings was circulated to all members of the forum for their information and each organization had the opportunity to include items on the agenda.

The relationship between the forum and the county council is illustrated in Figure 12. This structure of communication and influence was established in order to introduce relatively formal lines of responsibility and reporting to what otherwise might have proven to be a chaotic informal talking shop. However, this structure did cause subsequent problems in terms of ownership, which will be discussed in the next section.

The Lancashire community

Constituents of organizations

Lancashire Environment Forum

Forum Officers Steering Group

Environment Unit

Policy and Resources Committee and Planning Committee

Figure 12 The administrative framework co-ordinating Lancashire's approach

The LEAP process was one of the first initiatives in the UK (and indeed the World) to apply the principles of sustainable development through a Local Agenda 21 exercise. The key features of the process have been:

- the inclusion of a relatively broad range of interests in the decision-making process leading to the preparation of the action programme;
- the use of a consensus-building exercise to reach agreement on the key issues and proposed actions;
- the relatively informal and co-operative nature of many of the meetings and events leading-up to the final document; and
- the action-orientated emphasis of the process and the final programme.

Evaluating the LEAP experience

In evaluating the work of the forum in preparing the LEAP we have used Acland's framework as an initial starting point. However, we later expand our considerations to place this kind of policy-making process in its wider politico-economic context. The main issues that arise from Acland's ideal-type can be evaluated according to the six stages he outlined.

Assessment and preparation

As the first step in the four-stage process, the development of the green audit provided an important baseline of information on which to develop the action programme. There is a wide acceptance by the forum members of the green audit as the key source of information for LEAP, in part, because the forum members were the main information gatherers. Effective participative techniques were used in the consultation process. This enabled members to check the factual accuracy of the green audit and allowed both the public and forum members to be involved in the design of LEAP, by prioritizing the issues to be addressed.

The issues covered by the green audit recognize the cross-sectoral requirements of the new environmental agenda, with economic and health issues being raised alongside environmental considerations. Lancashire County Council's Environment Unit was responsible, however, for assimilating the information and drawing out the key issues. The forum, therefore, was rather more indirectly involved in the identification of the key issues than would be expected in a consensus-building process.

The voluntary approach to membership was successful in attracting the involvement and interest of the key sectors. Although the general high level of interest in the environment in the late 1980s and early 1990s contributed to the involvement of all the key sectors, the success clearly illustrates a demand for greater participation in environmental decision-making. The one sector weakly represented is that of industry. Although seen to represent a substantial proportion of the membership in Figure 10, the bulk of the participants are representative organizations, such as chambers of commerce, rather than individual businesses. Individual businesses are a vital instrument for implementation and without their involvement from the beginning some members have questioned the credibility of the process.

From the early stages of the process business interests perceived the county council as having a strong environmental and anti-industry stance. This led to a feeling that they were not being regarded as equal partners. As a result, the forum was not regarded as a high priority even by those companies who had their own environmental initiatives and were keen to improve their environmental performance.

Central-government organizations were well represented on the forum, although they tended not to take an overly active role. The Department of the Environment and the Department of Transport both made clear their observer status, while the Ministry of Agriculture Fisheries and Food and the Department of Energy (now incorporated into the Department of Trade and Industry) limited their involvement to one of ensuring that debate was well informed, rather than taking a proactive role.

The interest-group sector would also seem to be well represented on the forum. However, as much as 70 per cent of the interest groups are organizations which are nationally or regionally based (such as the Royal Society for the Protection of Birds and Friends of the Earth) and only 30 per cent are local organizations. As a result, the interest-group sector is dominated by what can be regarded as the 'institutional' voluntary sector.

The design of the structure of the process is one of the key areas of success. The co-ordination of over eighty organizations is very difficult, yet the structure of the specialist working groups co-ordinated by a steering committee has formed a useful framework. The selection criteria for the OSG has, however, resulted in a large number of local-authority representatives and in particular representatives of the county council. Although it is important to have interested parties involved in the steering committee, a poor representation of all sectors is a concern given the considerable responsibilities and influence of the OSG. Another feature is the low representation of specifically environmental interest groups (three out of twenty-three OSG members). This might seem rather ironic given the previously mentioned perception of the county council as pro-environment and anti-industry.

Initiation

As the lead organization in the process, Lancashire County Council has always been faced with the problem of ownership. For the process to have credibility throughout the community it should not be regarded as a county council initiative. This has proved to be especially important given the low opinion of local and central government held by local residents which was exposed by a social survey undertaken by Lancaster University (Lancashire County Council 1995) Although the problem has been recognized (e.g. the launch of the programme was not at the county hall), the process has had difficulty in removing the perception that the county council owns the process. Industry in particular has been suspicious of the lead role of the county council. Both the green audit and the subsequent LEAP have been perceived as county council documents.

Paradoxically, without the strong political leadership of the county council the whole process would probably not have got off the ground. The considerable political support ensured that resources were made available for the county council to service the forum. Similarly, as the county council is a key implementing agency and an important user of resources itself, it was necessary that the political decision-makers agreed with the objectives of the process.

Financial support for the forum has always been a problem. Public-spending constraints and the effects of recession on business, combined with the non-availability of funds in the voluntary sector, have resulted in a continuing dependence on the county council to service and organize the forum. Such a position of dependence does nothing to alleviate the problems of ownership and dependence faced by the forum.

With respect to the timing of the process, the forum has managed to keep to the schedules agreed and a remarkable amount of work has been achieved in a relatively short time. At the level of individual meetings, however, there has been less success. All meetings are held during normal working hours and therefore not all the groups who would like to be present have been able to attend. The local-interest groups have suffered in particular, with the voluntary representatives finding it difficult enough to allocate time to their own voluntary work, without committing more time to be represented at forum meetings.

Similarly, the councillor representatives of local authorities and other organizations have found it difficult to attend meetings unless they are already retired or have very sympathetic employers. There is a concern that existing links between organizations are simply being strengthened rather than encouraging new ones and that only the institutional voluntary sector is being represented with a more limited input from voluntary groups at the local level. However, the involvement of local interests in county-wide Local Agenda 21 exercises has been difficult in many other parts of the country given the tendency for these fora to concentrate on issues of a strategic nature.

Few of the organizations contributing to the LEAP process have formal or well-developed informal feedback mechanisms for relaying information and experiences back to their constituents. It would appear that at no point in the process have participants been made to agree or demonstrate how they propose to keep their constituents informed. As a result, although individuals may agree on an issue, there may only be a limited effect on the functioning of the organization that a participant is representing. The problem of poor feedback mechanisms is further compounded by the changing nature of organization representatives. Some of the larger organizations, in particular the national-government agencies, do not have a regular representative who contributes to the process.

Even the county council has not been free of these problems of co-ordination and feedback. In the key area of strategic land-use planning, for instance, there has been no formal feedback mechanisms established between the forum and the unit preparing the county structure plan. However, informal communication and the internal checking of the land-use implications of the LEAP has meant that the new structure plan (subtitled 'Greening the Red Rose County') has incorporated, or at least reflected, many of the ideas propagated in the forum. One important exception was the rejection by the structure-plan team of the LEAP proposal for tightening-up parking restrictions in the county's town centres on the grounds that it was unrealistic and inappropriate.

Negotiation: Exploration and exchange

The majority of the participants have responded positively to the process of consensus-building used for negotiation, information exchange and the exploration of ideas in the SWGs. Participants generally found the process very constructive and many even enjoyed taking part. Many were surprised at the amount of common ground which was identified during the discussions and the level of awareness and understanding of differing views that could be gained.

Perhaps inevitably some have suggested that the process involved too much talking and not enough action and, in particular, those involved in formal politics found the process to be rather slow. Although perhaps valid, such comments are inevitable due to unfamiliarity with the process of consensus-building. Individuals and organizations are more used to working within existing hierarchical structures, with which they are familiar and have had success in the past. As with anything new, there are likely to be those who are not readily convinced.

In Lancashire, the quality of the consensus gained by the SWGs varied between the four groups, with some groups producing much more challenging and innovative ideas. The variations reflected the different contributions made by individuals and the ability of each participant to adapt to the consensus approach. The contribution of individuals clearly influences the quality of the consensus achieved. The process ensures, however, that if nothing else, the less innovative groups did produce a level of agreement and the participants were better informed about some of the issues.

The role of the mediators, or lack of them, also played an important part in the quality of the consensus. Mediators should have a key role in ensuring that the generation and evaluation of options works to its full potential. However, the environment-unit officers, who were forced to take this role, were perceived as having an implicit county council bias. Thus, the SWGs lacked the direction and independence that a non-partisan mediator could have offered. In the context of the complex range of interests involved in Local Agenda 21, this casts some doubt on the ability of local-government officers to develop a role as 'empowering professionals' as suggested by Forrester (1989).

Decision-making

Once the options have been discussed, there comes a point where commitment by the participants is required. The use of the interactive day to test the draft proposals in workshop sessions was a good participative technique, which enabled all members of the forum to contribute to the development of the draft programme. The success of the process can be tested to some extent by examining the response received from the consultation exercise which followed.

Although the response to the consultation exercise was limited, with sixty-five of the eighty-two members not responding, this could be regarded as a positive indication of the success of the process. Indeed, many of the representatives of the organizations interviewed had been continually involved in the process of consensus-building and felt that no further comments were required. The responses that were received were categorized by the environment unit and are illustrated in Table 20.

As much as 24 per cent of the responses were, however, an outright rejection, which indicates that the early consensus-building approaches were not entirely successful. A major factor contributing to the rejections was the inability to get all forum members contributing equally and being represented on the SWGs. However, the fact that some of the rejections were based on the level of detail contained in the draft programme is an indication that the process has clearly produced something more than bland platitudes. Furthermore, considering that

Table 20 Draft LEAP consultation response

Response	Total number	%
Support proposal outright	57	22
Seek clarification	13	5
Observation	43	17
Suggested modification	83	32
Outright rejection	62	24
Total	258	100

40 per cent of the proposals were not commented on at all, and that out of the responses received over 40 per cent were points of observation, clarification or outright support, the draft programme has achieved a good level of consensus between the diversity of forum members.

The level of consensus achieved has, however, been variable. Although unanimous agreement has been achieved for a number of the proposals, a significant number do not attain such a high level of support. Many of the forum members have their own environmental initiatives and agree to a LEAP proposal only if the proposal does not contradict or cause problems for their existing policies. Furthermore, in some cases the OSG has used a definition of consensus which is based on at least 50 per cent of the members agreeing to an issue or proposal. Although not always the case, a decision could be taken with quite a low level of agreement, especially if the OSG is keen to promote it.

The need for practical and implementable proposals was recognized within the decision-making process. The proposals were 'reality tested' by examining the level of funding required and the organizations which might be involved in implementing them. Dealing with the financial issues, however, was clearly complex and was only considered to a limited extent. The issue of responsibility was addressed more directly, however, with the implementing agencies clearly identified next to each proposal in the final document.

The principles of consensus-building suggest that all participants must be prepared to put their name to the document. If any participant is unwilling to sign up, then the issues in question should be subject to further negotiation. However, in Lancashire, the items which could not be agreed have been included in the final text of LEAP and mechanisms have been established to tackle these disputes as part of an on-going process. This reflects the strong desire of forum members to move from talk into action as quickly as possible.

Although the county council have maintained a high level of corporate commitment to the principles of sustainable development and have responded to the LEAP throughout the organization, there are examples of disagreement between the programme and other policy guidance produced by the council. As mentioned earlier, the new county structure plan (in draft form) took a different view on the LEAP's emphasis on restraining car parking in the county's town centres. The planners involved in its preparation saw the structure plan as the primary document, with the LEAP acting as supporting and, where conflicts exist, subservient guidance. The same view of LEAP as a useful but subsidiary form of policy guidance was evidenced in interviews with officers from the district councils.

Maintenance

Contexts and circumstances change and a number of robust review structures have been set up in Lancashire which recognize the need to monitor change and adapt the response as required. In the case of the deferred issues, for example, opportunities for re-negotiation have been organized. To date, the informally constituted framework of the forum has provided the necessary organizational context for the process. However, there are always threats to the maintenance of this achievement and difficulties have been experienced during the last few years as a consequence of the local-government re-organization process and initial problems and delays with implementation.

Local-government review in Lancashire caused uncertainty during the last years of the Conservative Government and this is being prolonged by the new Labour administration's proposals for regional development agencies and regional assemblies. The informal nature of the forum has been satisfactory so far, given the political support from the county. If this support was weakened through local-government re-organization or regionalism, however, the forum may no longer have the strength to survive without some kind of formal status or organizational backing. Furthermore, the uncertainty surrounding local-government reorganization and regional government has pushed Local Agenda 21 activity further down the list of priorities for the majority of district authorities.

A legally constituted format is also useful for the implementation stage of the process, to attract funding or to lobby central government. This has been provided to some extent by the establishment of a number of centres for environmental excellence (CEEs) which were proposed in the LEAP. The most advanced centre is the one specializing in industry. The centre's key aim is to reach the under-represented sector of industry. Given that industry needs to see some kind of short-term benefit, and given that they are generally suspicious of long-term strategies, this may be the only realistic way to encourage industry to become equal partners. Another five CEEs were established in the early 1990s, whilst a further three have recently been set-up. The initial six centres placed a bid for funding under the European Union's LIFE programme during 1995/6.

Despite the important role of the CEEs, the implementation of other proposals in LEAP has been variable. One problem is the fluctuating level of interest in the environment since the initial upsurge of the late 1980s. This is compounded by the fact that some participants feel that the production of the LEAP document is an end in itself rather than being just the beginning of the process. It has been difficult to retain the commitment of participants after the production of the final document because there are no clear objectives for the process. As an illustration, the third interactive day, which assessed how best to take the process forward, was characterized by very poor attendance and, similarly, the response rate to the annual review questionnaire has often been slow. The production of the second green audit (in 1997) and the subsequent review of LEAP has tried to re-kindle the energy that was in evidence during the production of the original document.

Since the forum has no power to impose the agreed consensus and force action, implementation is inevitably difficult. Action has also been constrained during a period of financial stringency caused by government policy and economic recession. The programme has, however, only been in place for a relatively short period so it is still too early to really judge its overall impact. The process in Lancashire has been well structured and its incremental nature has ensured that the forum have not taken on too much at once. A number of weaknesses have been identified in this section although, in general, the practice in Lancashire has a close affinity with the principles developed by Acland and, as such, can be regarded as a moderately successful consensus-building process.

Consensus and economic interests

One of the clear lessons coming from the LEAP example is the difficulty of integrating key economic interests into the consensus-building process at the local level. Despite the rhetoric of economic partnership which pervades the statements of government policy and the pages of various economic and policy journals, there appears to be little evidence, either in Lancashire or elsewhere, that business involvement in Local Agenda 21 has been substantial or comprehensive. This is worrying given the whole emphasis of Agenda 21 (UNCED 1992) to expand the input of business and community interests in this global process of sustainable development.

Many writers have sought to explain this, and other, feature(s) of 'public' policy using ideas about 'policy processes' (Healey 1990) and 'policy networks' (Marsh and Rhodes 1992). This puts emphasis on the fragmented nature of state institutions and the varied avenues through which policy is negotiated, agreed and implemented. In both cases a typology of possible avenues has been constructed and these are briefly outlined in Tables 21 and 22, together with their application to the evidence from Lancashire.

LEAP fits a little uncomfortably into these frameworks and this, in itself, is interesting. It aspires towards open democratic debate and an open-issue network but has been modified through various pressures into a more closed decision-making process, at least at certain stages and through certain sub-organizational arrangements. An example of this is the particular role of the OSG at the core of the policy network. Here we find a relatively close set of inter-personal linkages where resources are exchanged and the agenda is set for the rest of the network. At the periphery, in the wider forum, minutes of meetings are received for comment and a process of consultation, rather than participation, is progressed. Furthermore, as the Lancashire Environmental Forum begins to implement its action programme, subtle changes are taking place in the composition of interests involved. Bargaining between the key implementation agencies appears to be replacing the more open debate that characterized the production of LEAP.

Another implication of using these (related) frameworks is that it draws our attention to other possibilities. Thus, just because business interests are not engaging enthusiastically with the LEAP process, this does not mean that they are excluding themselves from influencing environmental policy. Indeed there is evidence from the interviews in Lancashire to show that business interests feel they can be most effective at furthering their particular interests in this field by interacting with policy-makers at more regional or national levels. On a day-to-day basis, it comes down to asking themselves whether it is worth the effort. The Centre for Environmental Excellence for Industry appears to provide distinct (commercial?) benefits, whilst the talking-shop of the Environmental Forum was

Table 21 A typology of policy processes and the Lancashire case

Policy process	Policy and action determined by ...	Lancashire case
Client list	... the use of patronage in return for direct political support.	No evidence of this. Could creep in during decision-making on details of project beneficiaries from county council initiatives.
Politico-rational	... judgement of politicians in the formal arenas of representative democracy.	County council committees (Policy and Resources and Planning) formally 'approved' LEAP and politicians chaired the 'set piece' interactive days. However, most of the decisions were taken out-side the formal arenas of representative democracy, albeit with political support.
Pluralist politics	... political debate among pressure groups with politicians determining the balance of advantage in political terms.	Regionally constituted press-ure groups were influential in the debating process but politicians did not mediate the outcomes to any signi-ficant extent.
Open democratic	... open 'rational' debate, where all affected parties discuss the advantages and disadvantages of particular courses of action and reach agreement without domination.	The policy process most clearly sought after in the LEAP process; but not all interests represented or involved at all stages. Lancashire County Council found it difficult to re-linquish 'bureaucratic' or 'politico-rational' ownership and business interests suspicious of this.
Bargaining	... negotiation with specific groups over a specific issue where mutual dependency between the group and the state is involved.	An increasing feature as LEAP moves into implementation stage.
Special committees	... discussion and debate by a selected group of experts with a relatively well-defined agenda.	LEAP rejected the tight selection of key interests and agenda items; features which have characterized some attempts to develop 'specialist' components of LA21 in other parts of the country.

(*continued*)

Table 21 (cont.)

Policy process	Policy and action determined by ...	Lancashire case
Corporatist	... negotiation of a wide range of issues over a long period with specified representatives of specified ('functionally' defined) organizations	Corporatist negotiation not accepted as appropriate policy process for LA21. However, the Officer Working Group is dominated by functional interests and there are other important 'alternative' avenues for them to influence economic, development and environmental policies at the local, regional and national level.
Bureaucratic/ Legal	... the application of formal procedural and legal rules.	Key role of county council officers and committees instils a residual 'bureaucratic' flavour to the LEAP process.
Judicial/semi-judicial	... formal hearings which consider the arguments of conflicting interests with an assessor balancing their relative merits.	Not part of the LEAP process; but could be a significant influence on the 'feedback' to LEAP from planning inquiries and the examination in public of the Lancashire County Structure Plan.
Techno-rational	... the judgement of experts and scientific reasoning.	Influenced the LEAP process, but not a key feature. Possibly more important during implementation (e.g. the rationale and requirements developed for the centres of excellence).
Market-rational	... market of quasi-market mechanisms using the principles of supply and demand.	Playing an increasingly important role in implementation; despite the rhetoric of demand management under-pinning the new agenda.

Source: based on Healey (1990) and Healey et al. (1988).

less useful. This cautious approach will only be modified if institutional power is seen to be important at the level of the county; at present there is little evidence of this. However, power is a dynamic entity and if important economic policy decisions are transferred to more local levels of government or is made more accessible to them, then the pendulum could be swayed.

Currently there appears to be some moves to allow this shift in power. Three aspects are worthy of mention: the encouragement of a plan-led system of decision-making for planning; the weight now being placed on locally generated

Table 22 A typology of policy networks and the Lancashire case

Network criteria	Policy community	Issue network	Lancashire network
No. of participants	Limited number with some conscious exclusion	Large	Large number with no conscious exclusion
Type of interest	Professional/occupancy of a senior position	Encompasses a wide range of interest groups	Encompasses a wide range of interest groups, but limited local and business-interest involvement
Continuity	Consistent membership and values	Fluctuating access	Consistent membership with some degree of fluctuation, particularly of individual representatives
Frequency of interaction	Frequent/high quality	Contacts fluctuate in quality and frequency	Infrequent interaction but well structured
Consensus	Participants share basic values	A degree of agreement but conflict present	Initial conflict but a level of consensus achieved
Nature of relationship	All participants have resources, relationship is one of exchange	Some participants have resources, but limited	All participants have some resources; relationship is one of exchange
Nature of interaction	Negotiation over direction of policy	Consultation, with limited input to policy outcome	Negotiation over direction of policy
Power	Balance among members, mutual expansion of power	Unequal power, winners and losers	Balance among members; mutual expansion of power (at least during policy-making)
Structure of participating organizations	Membership will accept agreement	Variable capacity to regulate members	Variable capacity to regulate members

Source: based on Marsh and Rhodes (1992).

strategies for major central government and European funding programmes; and the new Labour Government's proposals for regional development agencies. It has to be stressed that these trends are fraught with the usual problems of inconsistent application; central-government retrenchment, caution or apathy; and the ever-present tendency to ad-hocism and opportunism which can undermine agreed policy. The further potential problem is that non-economic (e.g. environmental) interests could be excluded from economic policy debates, particularly at the regional level. However, there is little doubt that these movements in policy and structure offer the potential for local and/or environmental policy networks to increase their role and influence at some expense to more established economic policy communities at higher levels of the state apparatus.

Conclusion

What does this package of analysis and contemplation mean for the process of consensus-building and its contribution to sustainable forms of economic and social development? First, it is obvious from the LEAP example that particular care needs to be taken in devising an approach which is effective in building a locally agreed agenda. In particular it needs to give attention to all the dimensions raised by Acland. In the case of LEAP, the lack of independent mediators meant that the county council found it difficult to distance itself from the ownership of the process. This was especially unfortunate given its perceived anti-industry stance in the eyes of the business community. Other local councils (e.g. Gloucestershire and Reading) have benefited from the use of independent facilitators in the development of their Local Agenda 21 fora.

Second, in order to make policy and implementation processes meaningful and worthwhile there needs to be a measure of subsidiarity in resource and power terms which attracts the necessary stakeholders into the consensus-building process. This appears to have been a reason for the low input from businesses in Lancashire. Similar attitudes pervade the development industry (Carter and Darlow 1997), although other chapters in this book suggest that much can be achieved given the right conditions. Local-government reorganization is likely to have mixed impacts in this respect. In areas where unitary authorities are being established the concentration of local-government service provision will enhance the potential influence of environmental fora on certain key resources and powers. However, it will also dissipate the energy of local and regional organizations who may be asked to get involved in numerous local fora, rather than a single county-wide or regional forum. Balancing-up these forces, maybe it is at the regional level that the best opportunity exists to marry the globalism of economic and environmental processes with the localism of meaningful participation by ordinary people.

Finally, effective consensus-building (at whatever spatial level) requires an institutional design which prioritizes open debate and broad policy networking at the expense of more closed types of policy processes and networks. If these alternative policy processes are available for certain interests and state agencies to bypass public involvement in decision-making and possibly undermine those agreements then the confidence in, and commitment to the consensus-building process will ebb away. This requirement is, of course, closely linked to the second point above; without some significant elements of local autonomy there is little incentive to seek common agreement over the use of locally determined powers and resources. Ironically, one of the reasons for the development of locally based corporatism in the past has been the important role of local government in certain key economic-development activities, such as land-use planning, employment training, industrial development and transport (Healey *et al.* 1988; Simmie and King 1990).

Despite the very cautious set of conclusions listed above, there is no doubt that the central rationale of sustainable development continues to offer hope for the growth of broad, participatory policy networks. Thus, the interrelated set of environmental problems we all face means that any secondary conflicts of interest should, in the final instance, be subsumed under the broad umbrella of consensus necessary to tackle them. Following this line of argument, it is in everyone's best interest to promote sustainable development and the search for consensus that this entails. Whether this can work in the first instance in the various fora being developed around the world is still uncertain!

References

Acland, A. F. (1993). *Consensus Building, Reaching Agreement in Complex Situations*, London: The Environment Council.

Carter, N. and Darlow, A. (1997) 'Local Agenda 21 and developers: are we better equipped to build a consensus in the 1990s?', *Planning Practice and Research* 12(1).

Forrester, J. (1989) *Planning in the Face of Power*, Berkeley: University of California Press.

Friends of the Earth (1994) 'Planning for the Planet: Sustainable Development Policies for Local and Strategic Plans', London: FoE Joe.

Hardy, D. (1990) 'Consensus planning; realistic goal or idealistic dream?', *Planning Practice and Research* 5(2).

Healey, P., McNamara, P., Elson, M. and Doak, A. (1988) *Land Use Planning and the Mediation of Urban Change: The British Planning System in Practice*, Cambridge: Cambridge University Press.

Healey, P. (1990) 'Policy processes in planning', *Policy and Politics* 18(1).

ICLEI (International Council for Local Environmental Initiatives) (1995) *Case Study Series*, Toronto: International Council for Local Environmental Initiatives.

Lancashire County Council (1995) *Public Perceptions and Sustainability in Lancashire: Indicators, Institutions and Participation*, Preston: Lancashire County Council.

Marsh, D. and Rhodes, R. A. W. (eds) (1992) *Policy Networks in British Government*, Oxford: Oxford University Press.

Marshall, T. (1994) 'British planning and the new environmentalism', *Planning Practice and Research* 9(1).

Shorten, J. (1993) 'Environmentalism and sustainable development in contemporary planning practice', MSc dissertation, School of Planning Studies, University of Reading.

Simmie, J. and King, R. (eds) (1990) *The State in Action: Public Policy and Politics*, Pinter.

UNCED (United Nations Conference on Environment and Development) (1992) *Agenda 21*, New York: United Nations.

15 Local Agenda 21 and economic development

A case study of five local authorities

Dean Patton and Ian Worthington

Introduction

It has become increasingly recognized that effective responses to the environmental challenge demand action by all sections of the community and that much of this action must occur at the local level and involve local government. Following the Rio Earth Summit in 1992, local authorities have become the focus of an attempt to create strategies for 'sustainable development' under the Local Agenda 21 initiative (LA21), endorsed by national governments. Central to this initiative was the proposition that local authorities should create a 'partnership' with representatives of local businesses, voluntary organizations and the local community in order to produce a consensus on sustainable development within their respective areas. This implies not only the need to raise awareness of sustainability issues, but also an acceptance that the formulation and implementation of effective sustainable development strategies requires the active involvement, support and co-operation of all sectors of society, including groups traditionally disadvantaged in the policy process.

In responding to the demands of this initiative, local authorities have been forced to consider how LA21 relates to their other responsibilities, particularly those associated with the encouragement of business activity and wealth creation. As both the European Commission's 5th Environmental Action Programme and the Rio accord have indicated, questions of environment and development are intertwined and accordingly local authorities need to find ways of reconciling local economic and environmental objectives and policies. Such a reconciliation requires not only a general commitment to the notion of sustainable development on the part of government at all spatial levels – backed up by adequate funding from the centre – but also a willingness to consider whether current structures, processes and competencies enhance or inhibit the capacity of local government to respond. Nowhere is this need for self-examination more apparent than in the demand that policy formulation and implementation should be based on processes of consultation, consensus and partnership-building within the local community.

Research aims and methodology

This chapter is based on a research project which began in 1994. The first stage of the project was an analysis of UK local-authority responses to the LA21 initiative

(Worthington and Patton 1995). Using a postal questionnaire sent to all county councils in England and Wales (47) and to all metropolitan authorities (68), the authors examined five key areas of policy development:

- local-authority awareness of LA21;
- information sources used;
- council responses to the initiative;
- consultation and partnership-building;
- funding.

The findings of this survey – which had a response rate of 63 per cent – clearly indicate that a number of problems have been encountered by the majority of authorities faced with developing an LA21 strategy.

Against the background of this baseline information, the second stage of the research examined in more detail the approach used by selected authorities, chosen because of their positive response to the demands of LA21 (Patton and Worthington 1996). On the basis of structured interviews with representatives of Croydon Borough Council, Gloucestershire County Council, Hertfordshire County Council, Leeds City Council and Sheffield City Council, this phase of the research provided an insight into how some authorities have sought to tackle the problems identified in the earlier study. The aim was not to suggest a framework through which local authorities could develop the initiative, since no single paradigm clearly exists. Rather it was to identify some of the factors which may assist the process of strategy formulation and implementation in an area fraught with conceptual as well as practical difficulties.

In carrying out the second stage of the research it became evident that in all the authorities questioned there was a clear recognition that the notion of sustainable development provided a basis for the integration of local economic development and environmental improvement. To investigate how far this perception had been translated into action, the authors then examined internal documentation and carried out a further telephone interview with the representatives from the structured sample. The primary aim of this part of the project was to ascertain how different local authorities were tackling the problem of reconciling the economy and the environment.

Phase 1: Identified problems

Funding for LA21

Only a quarter of authorities in the initial survey appeared to have earmarked funding specifically for LA21 and in many cases this funding was from existing budgets rather than being additional money allocated to the initiative. Perhaps not surprisingly, about half of all respondents claimed that funding devoted to LA21 was inadequate, whilst a further 20 per cent said it was too difficult to judge at the moment whether funds would prove adequate.

The establishment of formal structures

Arguably the most significant finding from the survey was the relative lack of progress made in some local authorities towards the establishment of a structure which was designed to co-ordinate the authority's response to LA21. At the time of the questionnaire less than half of all respondents claimed to have established a formal co-ordination mechanism for developing a local sustainable development strategy, although recent evidence suggests that UK local authorities are becoming more aware of the implications of LA21 at local level (see, e.g. Tuxworth 1994)

There are two areas of specific interest that arise from the questionnaire. First in relation to leadership, control of the initiative varies between authorities, but two models tended to predominate, each with attendant problems:

   i A minority of councils had designated an individual with overall responsibility for co-ordinating the authority's response to LA21 and for liaison with other local authorities and bodies at local, national and international level. This individual had normally been seconded from an existing department and her/his salary normally consumed the majority of any budget available. Development of the initiative was, therefore, dependent to some degree upon the ability of the individual to gain extra funding and support from the various departments and this had sometimes proved problematical when requests had been seen as part of an attempt at 'empire building'.
  ii Most authorities seemed to have located responsibility for LA21 within a specific department, under the purview of a lead officer or small group of officers. Whilst this arrangement was consistent with existing practice, it often resulted in the initiative being identified as essentially the concern of one department, with other departments taking the opportunity to distance themselves from its requirements, thus compromising the development of an integrated approach to sustainable development.

Second, in relation to location, where councils chose to locate the initiative within the existing departmental structure, there was little evidence of any consensus across authorities about which department should be responsible; over twenty-three different responses were cited. Whilst this diversity of approach had its merits, it threatened to limit the ability of local councils to liaise with and learn from one another, arguably a significant fact – given that the local-authority network appears to be a key source of information and advice (Worthington and Patton 1995). It could also reduce the possibility of collaborative projects across boundaries that could offer synergistic benefits.

Development of consultation and partnership building

Despite the exhortation that local authorities should involve local businesses, voluntary organizations and citizens in the formulation of an LA21 strategy, the evidence suggested that such involvement had been relatively limited. Only one-quarter of all respondents claimed to have had direct consultation with local businesses, whilst one-third had spoken with local voluntary organizations. In the case of local citizens, only one-sixth admitted to any face-to-face consultation, though a number suggested that such consultation was planned at some future date.

Lack of central-government support

Another finding from the original questionnaire was that central government, although quick to support the initiative in the form of rhetoric, had provided little support in terms of finance or direction. A typical response by those authorities that had done little more than test the water was to state that they intended to wait until further clarification and support was provided by central government.

Phase 1 summary

The first stage of the research project clearly indicated that there were a number of problems that an individual local authority had to overcome. Central-government policy is clearly outside of a local authority's direct control and the provision of funding is restricted by central-government fiscal policy and by the various competing needs upon very tight budgets. The development of formal structures and methods by which local participation could be undertaken, however, is within its sphere of influence and whilst there are various means through which participation and consultation can be encouraged, it became clear that many authorities had yet to find approaches that involved and integrated members, officers, business and local interests. The majority of authorities still appeared to be persevering with the normal patterns of consultation through the usual channels of communication. Accordingly, 'top-down' approaches still tended to predominate and this evidently hindered the creation of a true partnership which could function at all stages of the policy process, from ideas generation to the formulation and implementation of an LA21 strategy.

Phase 2: Practical approaches – examples from five local authorities

Every local authority has its own way of responding to the problems identified above and its response will be conditioned by a variety of factors including finance, political complexion, precedence, culture and personnel. In the analysis below, the approach of five councils is examined in some detail.

Funding for LA21

It is widely accepted that there is no simple solution to this problem. As with other initiatives imposed on local government by the centre, it is often a matter of finding ways of redirecting money from existing budgets and/or of persuading others (e.g. the private sector) to bear some of the cost. The success of either of these approaches appears to be conditioned in part by attitudes to environmental issues within each local authority and in its wider community.

It was noticeable that within the authorities interviewed, environmental concerns tended to have a high profile, largely because of the activities of individuals and/or groups who had championed the cause for some time. Each of the five authorities already had a history of positive involvement in environmental matters and this had helped to raise awareness and create support for actions designed to improve the environment. To this extent, LA21 was viewed less as a 'threat' or 'problem' and more as a continuation of existing policies. In effect it reinforced the existing culture of the organization rather than challenged it.

Leeds provides a good example of how initiatives can be incorporated into existing activities. As one of the four 'environment cities' in the UK, Leeds had been able to subsume LA21 under the environment-city umbrella and this had offered significant scope for funds to be used for activities which are seen as meeting both purposes. Some of the sustainable development projects were regarded as departmental responsibilities and were funded from existing departmental budgets; others under the environment-city programme attracted financial support from other organizations in the public and private sector, thus reducing the burden on the city council.

Hertfordshire had benefited in a similar way from its designation as one of the pilot authorities on the National Sustainability Indicators Project, with some additional funding available for joint research projects, such as its investigations into soils and sustainability – jointly funded with the Department of the Environment and Bedfordshire County Council. Sheffield, in contrast, had encouraged a community-wide commitment to environmental good practice which had been reflected in departmental budgets and had helped to generate a modest amount of corporate business support.

The establishment of formal structures

If local government is to work towards more sustainable forms of development it is important that LA21 commands widespread support across local authorities and that its principles help to guide those involved in the decision-making process. The creation of structures which help to encourage such a cultural change is therefore to be seen as a means to an end, not an end in itself.

Whatever structure is chosen, it should be capable of generating co-operation and support for the initiative and should encourage integration within and between authorities. Simultaneously it should provide for involvement by representatives of local interest and should provide a means through which the local community can participate in the generation of ideas.

Within the structured sample it is interesting to note that all five authorities questioned appeared to have found different ways of achieving most, if not all, of these goals. At Croydon, the council had established an LA21 members' sub-committee, supported by an officers' environment working group and by a number of project groups with cross-departmental representation and local-community involvement. Hertfordshire had a green-issues panel within the full council and used both a structure-plan policy panel and a county-wide environmental forum, with the latter having a members' steering group and officer working groups and incorporating the ten district councils and the Hertfordshire Association of Local Councils, along with a number of outside interests. Sheffield's arrangements included a departmentally-based environmental working party – which was a sub-part of the Council's Policy Committee – and an environmental forum representing local interests. The council also relied considerably on one individual to co-ordinate the various elements in a people-focused initiative known as the 'living city', rather than LA21.

In Leeds, the local authority had developed a variety of structures linked together through the council's 'green strategy', which was launched in 1991. Every council department was required to develop an environmental action plan and was linked at political level through the Green Strategy Steering Group which considered issues which had a corporate dimension. The environmental work of departments fed into the environment-city initiative which included the authority's response to LA21. Well-established fora of local interests (see below) and a network of informal as well as formal relationships provided for broad participation by the different vested interests.

Local consultation and partnership-building

In its Local Agenda 21 Roundtable Guidance notes to local authorities, the Local Government Management Board stresses the importance of all agencies at local level working together to prepare, promote and act on a shared LA21 (LGMB 1994). This is an area where the five authorities had been particularly active.

Prior to its campaign for environment-city status Leeds had established the Leeds Environmental Business Forum (LEBF) and the Leeds Environmental Action Forum (LEAF) in May 1992. At the time of the interview the LEBF had 115 member companies – predominantly small and medium-sized enterprises (SMEs) – and offered its members a number of services including an initial environmental review and guides to environmental good practice. As an arena for discussion on local issues, the forum has consistently provided a mechanism for pooling expertise and knowledge and for incorporating the business community into the decision-making process. Similarly with LEAF – which includes representatives of community groups, charities, voluntary organizations and individuals – the forum served to bring together members of the local authority and the wider community and provided a network of organizations whose accumulated voice had significant power in the local area.

Sheffield's Environmental Forum, established in 1990, was a partnership of organizations, groups and individuals committed to the conservation of the city's environment and it worked closely with the city council, having three representatives on the Environmental Working Party. In addition the city had also established a green business club which brings together the city council, local businesses, the universities and a number of other representative bodies. Hertfordshire's Environmental Forum, also established in 1990, provided for links between the various local authorities within the county and local interests, including businesses and local citizens – with 'roundtables' and 'focus groups' a favoured mechanism for local involvement.

Croydon had sought to encourage participation by incorporating outside interests – including Friends of the Earth, local-residents' associations and the chambers of commerce – into the formal structure of established project groups and as such had achieved a regular, if somewhat limited, degree of local involvement. Gloucestershire, by contrast, had adopted the novel solution of developing the LA21 initiative within the local community by using an established local environmental trust known as the Rendezvous Society which was grant-aided by the county council. Through a complex system of conferences, working parties, focus groups and a co-ordination mechanism, the local authority had achieved substantial participation by a wide cross-section of local stakeholder groups. In total it was estimated that there were around 500 actively involved in the different activities, including the environmentalist Jonathan Porritt and senior personnel from the large privatized utilities and the former National Rivers Authority.

Lack of central-government support

It is no secret that central government has been less than wholehearted in its support for the LA21 initiative and as a number of commentators have pointed out this has not made local government's job any easier (see, e.g. Harris 1994). Local Agenda 21 is not a substitute for national policy action on the environment, it is a complement to it. Integral to the success of the initiative are the actions of the national government; it is at this level where many of the sustainability issues must first be addressed (Levett 1994).

Like the other authorities in the original sample, the five interviewed authorities recognized the role that central government can play in encouraging individuals to choose more sustainable lifestyles and in removing some of the restrictions on local government which would allow councils to be more effective in creating policies and programmes which benefit the environment. One area of particular concern was the tendency of central government to direct funding and responsibility towards quangos and away from local authorities, thus potentially hampering the development of local sustainable development strategies. If each local authority is to achieve its target date of 1996 for reaching a consensus on an LA21 strategy, this is one of many issues which central government needs to address urgently.

Summary of Phase 2

All five authorities interviewed suggested that at least some of the progress which had been made on LA21 was due to influences which pre-date the Rio accord and, in particular, to the activities of individuals and groups who had been active on environmental issues. This activity had clearly accentuated public awareness and had helped to raise the profile of sustainable development at both officer and member level. To this extent LA21 tended to be seen more as a continuation of existing policy and programme developments, rather than as a marked shift in policy direction or in underlying attitudes and values and this has undoubtedly been beneficial in building support for the initiative within each respective policy community.

In interviewing representatives of the sample authorities, one could not fail to recognize the degree of commitment and enthusiasm for an initiative which appeared to have a large measure of cross-authority support at both political and administrative level. At the same time, it needs to be said that this support was often conditional and was not necessarily shared within the wider community where the notion of sustainable development is not widely understood. Whilst the five authorities have gone some way to raising public awareness of sustainability issues and to involving local groups and individuals in the policy process, much remains to be done in this area if a true partnership of local interests is to be created. This is a challenge which faces all local authorities and which strikes at the very core of the traditional pattern of local democracy. It remains to be seen how far local authorities are permitted and/or prepared to go in responding to this challenge over the next few years.

Phase 3: Linking LA21 and economic development

The concept of 'sustainable development' requires local authorities to adopt an integrated approach to economic and environmental aims and policies between departments and across sectors (see, e.g. Gibbs 1991 and 1994; Jacobs and Stott 1992; Barrett 1995). In the UK existing statutory and administrative arrangements provide a general framework within which integration can be pursued. Apart from the provisions of the Local Government Housing Act (1989) regarding the formulation of local economic development plans, local authorities are required to address the question of environment and development in their UDPs and DDPs and to take account of governmental guidelines as laid down in the various planning-policy guidance notes. PPG4, for example, places emphasis on the need for development plans to take account of both the location demands of business and wider environmental objectives thus recognizing the importance of the planning system in integrating environment and economic demands. Similarly, PPG12 and 13 set out a framework of national (and regional) policy guidance designed to incorporate sustainable development concepts into planning policy (Gibbs 1994).

The ability and/or the willingness of local authorities to reconcile the demands of LA21 and economic development needs, however, to be seen in a wider context. LA21 is a relatively recent initiative and is a non-statutory undertaking, two factors which have limited the general response. Added to this, uncertainties caused by local-government restructuring and reform and perennial problems of funding have helped to divert attention away from less immediate issues. As far as the latter is concerned, local authorities have not only faced tight budgetary control – a factor which inevitably impacts upon their non-statutory obligations – but the funds which have been made available have as yet not been targeted at LA21 in particular, or sustainable development in general. As Levett (1994) has pointed out, local authorities need both the powers and resources to implement initiatives such as LA21 and these are best developed within the context of a regional and national commitment to sustainable forms of development. Without such a commitment from the centre it is questionable how far local authorities will be able to go in developing the kind of integrative approach to policy recommended by commentators such as Geddes (1993).

The private sector has also experienced difficulties in coming to terms with the need to link economic and environmental development. Although business is only one of the specified groups to be incorporated into the LA21 process, exactly how this is to be achieved and to what extent it is to be involved is left essentially to individual authorities. Whilst this *ad hoc* approach has its merits, it may not suit all types of organization, particularly larger enterprises with sites in different localities which tend to prefer a more-consistent policy on local involvement. Whether such consistency would be sufficient to encourage greater moves towards a more sustainable approach to development is, however, highly debatable. As a recent study of business developers and the LA21 process has shown, there is considerable reluctance on the part of many businesses to become involved in an initiative which is regarded as 'airy fairy', relatively unimportant and of little immediate benefit (Carter and Darlow 1996).

Given the context of LA21 within local authorities, the lack of adequate funding from the centre and the difficulties of engaging the private sector, one should not expect the integration of economic and environmental development to be too far advanced. Accordingly, the following evidence from the five sample authorities should be seen as an indication of the ways in which different councils have begun to tackle the question of reconciliation, not as a final solution to what is clearly a difficult problem. One of the keys to achieving more sustainable forms of development is for all sectors within the local community to accept that the integration of economic and environmental activities needs to be pursued and that incremental change is better than no change at all. As all of the questioned authorities have readily conceded, whilst a start has been made, much still remains to be done if sustainable development is ultimately to permeate all local-authority policies and programmes in the foreseeable future and to gain widespread support at community level.

Commitment

Judging by statements of intent, there is ample evidence to show that all five authorities questioned have a commitment to integrating environmental considerations into the economic-development process and this commitment is invariably incorporated in mission statements, corporate and/or unit business plans, and/or economic policy documents. Croydon's Economic Strategy 2000 Draft Action Plan for 1996/97, for instance, has as one of its key objectives the aim of, 'improving the quality of life and the environment' for local citizens through a series of initiatives which include projects which are central to its current work on LA21. Similarly in the authority's Consultation Draft (1995), alongside other key corporate objectives, numerous references are made to the need to link economy and environment in order to create 'an environmentally sustainable driven market economy'.

Leeds' green strategy adopts an equivalent philosophical stance in calling for a strategy of sustainable economic development which recognizes the need for economic growth and environmental protection to proceed in step. This corporate commitment to greening the authority's approach to the local economy is also echoed in Sheffield where the council is seeking to link economic and sustainable development approaches through a programme entitled 'Growing Together'. As in the case of the other three authorities, Leeds and Sheffield clearly accept that issues of economic regeneration and sustainability are part of the same agenda. The problem within each authority, therefore, is how to achieve a co-ordinated approach to policy development in a way which encourages community involvement in the decision-making process.

Structural developments

In addressing the question of policy formulation each authority seems to have evolved an approach which builds largely upon existing arrangements and which frequently contains individuals at administrative and/or political level who have shared responsibility for environmental and economic matters. Thus, in Hertfordshire, the policy panels for LA21 and economic development are housed administratively in the same office and are serviced by officers who work side by side and who report back to the same individual, although there is limited cross-over of staff between the two panels, members do sit on both panels. Croydon's system of LA21 project groups includes one entitled 'Work and Economy' which is chaired by the head of the Economic Development Programme and a new post has recently been created to deal with questions of environment and economic development within the LA21 process.

Gloucestershire and Leeds provide further examples of structural links between the two areas of responsibility. In the former, Vision 21 (the authority's LA21) and economic development activities are under the same management system and share some of the same personnel in a partly devolved decision-making process. In the latter, the integrating mechanism is the 'Leeds Initiative' which is a partnership scheme which brings together representatives of the public, private and voluntary sectors with the aim of developing and presenting Leeds as a 24-hour international city. Representatives of the Leeds Development Agency which runs the initiative are actively involved in both economic development activities and the LA21 process and provide members for the different specialist working groups that contribute on issues such as economy and the environment.

At a political level too, evidence exists of links between LA21 and economic-development activities, although these are not always as formal. In some councils, environment and economic-development committees or sub-committees share some of the same members (e.g. Sheffield) and these committees frequently report back to the same policy committee. As indicated above, the planning process requires elected representatives and their advisors to take account of environmental considerations in shaping development decisions. This require-ment seems destined to be an increasingly important influence on planning policy in future years.

Consultation and community participation

Local-authority planning procedures have traditionally included a 'top-down' consultative process in which local groups and individuals are asked to comment on planning proposals shaped by the local authority. In contrast, LA21 favours a more 'bottom-up' approach to decision-making with the local authority playing an active role in creating community-wide partnerships to design and implement strategies for local sustainable development. How have the sample authorities dealt with this apparently contradictory view of government/community relation-ships when seeking to reconcile economic development and environmental improvement?

Though it is perhaps too early to give a definite answer to this question, there is some evidence to suggest that increased links between the two activities have helped to encourage greater community involvement in economic-development decisions, particularly in cases where existing organizations provide a forum for stakeholder participation (e.g. LEAF and LEBF in Leeds, the Green Business Club in Sheffield). Whilst for the most part such involvement tends to centre around particular projects, in Gloucestershire housing the LA21 process in the local community appears to have influenced its approach to strategic planning. Later this year (1996) as part of its strategic-plan consultation process and LA21, the authority is planning to hold an economy stakeholders' conference at which groups and individuals can articulate their views on issues of economic and environmental policy. Other local authorities will no doubt watch this develop-ment with interest.

Funding for an integrated approach

Funding for the integration of LA21 and economic development at a local level has not been available thus far from any central source. Indeed funds for the development of LA21 are only provided at a local level by the respective authorities and remain discretionary. The failure to link funding either at a local or national level to a more integrated approach must be regarded as a major factor in limiting a more integrated environmental/economic approach to development. The arrival of the Single Regeneration Budget (SRB), however, has created a tenuous connection. The SRB has been developed since the Rio

Summit and the publication of the UK's Strategy for Sustainable Development. This is clearly reflected in the guidance notes which make reference to the strategy and LA21. Nevertheless, as Bennet and Patel (1995) have indicated, despite the rhetoric there is little evidence that the new government thinking on sustainability has been incorporated into the (SRB) programme. The main hope lies in the broader set of output measures that have been incorporated into the SRB than was previously available under the city-challenge initiative. Furthermore, within the SRB there is a recognition that the achievement of immediate short-term outputs may not necessarily lead to the achievement of long-term objectives in an area. In consequence the SRB may also be used to contribute to an overall strategy for a relatively wide range of issues and this may include LA21 strategies. It must be emphasized that this possibility will only materialize if local authorities choose to incorporate such measures within their individual bids.

Summary of Phase 3

Policy commitments entered into by governments at international, supra-national and national level have required local decision-makers to seek ways of reconciling their environmental and economic activities within a framework of cross-sector collaboration. As this research project has demonstrated some of the more proactive local authorities appear committed to, and in many cases have developed, a structure that is consistent with pursuing policies that approach the economy and the environment in an integrated manner and their experiences provide a useful source of information from which others can learn. There are, however, at least three major limiting factors which appear to affect the capacity of local government, generally, to respond and/or to assess how it has performed. First, the 'bottom-up' consultation process is atypical to normal local-authority practice. It engenders an *ad hoc* approach that may be useful, for example in the generation of ideas and the motivation of under-represented groups within any community, but which fails to allow for a more-strategic approach both within and between local authorities. It may well be the case that business involvement is needed at an executive level to convince others of the relevance and importance of the LA21 initiative and that a mixture of top-down and bottom-up approaches is likely to prove useful in this context.

Second, the pitiful level of central government support for LA21 and for the integration of economic and environmental programmes is clearly a significant factor in the limited response of local authorities. Current evidence suggests that the SRB may offer a method of funding certain projects identified under the LA21 initiative thereby allowing environmental considerations to be incorporated into economic-development strategies. To develop a more comprehensive coverage, however, more dedicated funding will be required which provides for a full range of sustainability objectives to be integrated into future economic-development plans and programmes.

Third, the ambiguous nature of 'sustainable development', coupled with the problem of reconciling the needs of the different stakeholder groups, has

254 *D. Patton and I. Worthington*

undoubtedly inhibited the development of evaluative tools by which local authorities are able to assess their progress in environmental terms. This problem is likely to prove important not only in the context of LA21 but also in the search for methodologies which support any integrated approach to budgetary, programme and project appraisal.

References

Barrett, I. (1995) 'How to green local economies', *Town and Country Planning*, March.
Bennet, J. and Patel, R. (1995) 'Sustainable regeneration strategies', *Local Economy* 10(2).
Carter, N. and Darlow, A. (1996) 'LA21 – none of our business?' *Town and Country Planning*, April.
Geddes, M. (1993) 'Local strategies for environmentally sustainable development', paper given at Urban Change and Conflict Conference, Sept.
Gibbs, D. (1991) 'Greening the local economy', *Local Economy*, 6(3).
Gibbs, D. (1994) 'Towards the sustainable city: Greening the local economy', *Town Planning Review*, 65(1).
Harris, T. (1994) 'International agenda, local initiative', *Town and Country Planning*, July/August.
Jacobs, M. and Stott, M. (1992) 'Sustainable development and the local economy', *Local Economy* 7(3).
Levett, R. (1994) 'Options from a menu', *Town and Country Planning*, July/August.
LGMB (Local Government Management Board) (1994) *Local Agenda 21 Roundtable Guidance: Community Participation in Local Agenda 21*, Luton: Local Government Management Board.
Patton, D. and Worthington, I. (1996) 'Developing Local Agenda 21: A case study of 5 local authorities in the U.K.', Sustainable Development 4.
Tuxworth, B. (1994) 'Blazing a Local Agenda 21 trail', *Town and Country Planning*, July/August.
Worthington, I. and Patton, D. (1995) 'Researching UK local authority responses to LA21', *Local Government Policy Making* 22(2).

Part V
Conclusion

16 Retrospect and prospect

Designing strategies for integrated economic development and environmental management

Peter Roberts and Andrew Gouldson

Introduction

A number of important themes have emerged from the various discussions and assessments that are contained within the chapters of this book. These themes both reflect the general topics which were outlined in Chapter 1, and also represent the various manifestations of the interaction between economic development and environmental management that are associated with individual policy regimes and jurisdictions. Thus, for example, whilst the analysis demonstrates the ways in which the general stance of economic and environmental policy has been adjusted so that it is more in accord with international and national policy developments, the specific response to the desire to integrate economic development with environmental management is also shaped by the preferences and procedures that are expressed by regional and local governments. This relationship between the general and the particular can be identified in many of the cases that are presented herein, and this analysis places particular emphasis on both the continuing importance of place and the frequently divergent roles that are performed by administrations operating at different levels within the hierarchy of government.

Some of the other important themes that have emerged in the chapters of this book include:

- the importance of the role performed by technical approaches and procedures in encouraging and ensuring the closer integration of economic and environmental concerns;
- the desirability of developing and evaluating the two strands of policy in parallel;
- the need for improved mechanisms for policy development and implementation;
- the benefits (and costs) that are associated with the promotion of policy integration;
- the wide variety of approaches and responses that operate under different policy regimes;
- the influence that is exerted in the integration of policy by political structures and a wide range of associated institutional relationships;

- the important role that is played by space and place in determining both the potential for a change in the direction of policy and the likelihood that such a change will be successful;
- the importance of encouraging and supporting transnational action;
- the wide range of policy instruments that exist and the often unexpected origins of these methods and approaches;
- the urgent need for the further development of theory and explanation in order to assist in the advancement of our understanding of how and why different regimes of policy and policy integration have emerged, and what is likely to occur in future.

The following sections of this chapter reflect the above themes, whilst also introducing additional material from the general debates on both policy integration, especially through the promotion of territorial integration, and a number of other topics that have provided the focal points of the book.

Much of the material that is presented in the following sections of this chapter can be linked back to the themes which were discussed in Chapter 1 and, in particular, to the construction and role of ecological modernization theory. As was discussed there, ecological modernization is generally considered to provide both a source of explanation of contemporary trends and a prescription for policy development and implementation. Whilst ecological modernization displays a number of distinguishing characteristics, it can also be considered as one amongst many attempts to explain the origins and outcomes of the various processes of policy transformation, integration and divergence that are evident in any consideration of the relationship between the economy and the environment. In addition, there are a number of critics who point to the limitations of ecological modernization, both as a theoretical explanation and in terms of its use as a policy tool (see, e.g. Spaargaren and Mol 1991). Alternative theoretical models exist – regulation theory, for example, offers considerable potential (Gibbs 1996) – as do a number of other approaches that seek to guide policy formulation and implementation (Woodgate 1997). However, despite its weaknesses, ecological modernization provides a useful portmanteau of theory and methodological prescription that can be applied to various policy systems at different stages in their development. There is still a clear need for the development of over-arching theory and for policy guidance that can help to explain and predict the nature and outcomes of the relationship between economic development and environmental management.

Key themes and conclusions

As was noted in the preceding section, a number of common themes can be identified in the discussions and case studies that have been presented in this book. Although the relative importance of any individual theme varies from place to place and may have increased or decreased in significance over time, a general trend can be identified that demonstrates the development of an agenda of issues that must be addressed if the process of policy integration is to be extended. Indeed, there is one school of opinion that holds the view that the economy–environment interface should not be considered separately from a broader discussion of sustainable development, and especially from the social justice and welfare dimensions of sustainable development. This is correct, but it does not diminish the validity of the stance adopted in this book, indeed it simply acts to sharpen the focus and points to the value of adopting a comprehensive and integrated view of the design and discharge of strategies for sustainable development.

A further general theme that is evident from the chapters contained in this book is the importance of designing strategies for integration that match the requirements and opportunities that are evident in individual places. People, activities and ecological phenomena exist in places not sectors, and the social values which are associated with the qualities of an individual place are often fiercely defended (Roberts 1995). This attachment of people to place is a characteristic that is increasingly recognized in the design of policy at all levels from the supra-national to the local. The recently developed European Spatial Development Perspective, for example, attempts to provide a Europe-wide perspective on the territorial development and management of various aspects of national, European Union and international policy, including the desirability of reconciling the requirements of economic progress and environmental enhancement (Ministers Responsible for Spatial Planning 1998). Similar strategies can be seen at various spatial scales in many countries and continents. Each EU member state, for example, has established its own sustainable-development strategy which, in turn, provides the framework for the development of regional and local strategies and of sectoral implementation programmes. This is subsidiarity in action, and it reflects the view that many elements of the sustainable-development agenda are best developed and implemented at a regional or local level (Cohen 1993). Furthermore, there is a growing body of evidence which suggests that an even more effective approach is to incorporate an environmental agenda within economic-development or community-regeneration programmes at the outset. The amalgamation of parallel strands of policy in order to establish a single territorial programme for the development and management of a region or locality is a logical step towards more effective integration.

The argument which has been introduced above is an important component of a much wider debate on matters such as subsidiarity and policy performance. The dimensions and outcomes of the first of these debates can be illustrated by reference to a number of the chapters in this book. For example, the New Zealand case demonstrates the various ways in which a local agenda for the integration of economic and environmental concerns can be developed within a national context that provides both encouragement and support for policy innovation. Equally, a number of the local case-study chapters from the UK – especially those concerned with the West Midlands, Leicestershire and Fife – demonstrate the importance of providing institutional 'space' within which appropriate structures and responses can be developed. The second issue – policy performance – is more complex and is difficult to illustrate by reference to a specific example. However, in a general sense the issue of policy performance is one which pervades many debates and it is an ever-present concern at all levels of government. Thus, for example, one of the key reasons advanced in support of the introduction of the environmental assessment and appraisal of programmes and projects is that it helps to prevent the waste of both financial and natural resources by eliminating unsound prospects at the outset or by allowing for the redesign of less effective proposals. Elizabeth Wilson has illustrated some of these concerns in her chapter, whilst other evidence of the search for greater policy efficiency can be seen in Chapter 2, which offers evidence from the EU of the attempt to develop policy vehicles that allow economic development to be integrated with environmental management in, for example, the regional development programmes that are supported by the Structural Funds.

This general discussion of some of the overall themes and issues that have been identified in this book provides the basis for the further consideration of a number of selected topics. These topics reflect the themes that were referred to in the first section of this chapter and they also attempt to anticipate the form and content of the future policy agenda; they include the following items:

- the identification of the lessons from existing practice that may be of assistance in other places and in the future;
- an appreciation of the improvements that can be achieved in terms of the techniques, methods and procedures employed for both policy development and implementation;
- an assessment of the relationship between the degree of policy integration that can be achieved and the quality of territorial governance and management;
- the implications of the division of responsibilities for policy formulation and implementation between the public, private and voluntary sectors;
- the extent of the influence exerted by global debates upon the form and content of policy at transnational, national, regional and local levels;
- an evaluation of the explanations which underpin our understanding of the interaction between economic development and environmental management and the adequacy or otherwise of current theory.

Lessons from practice

Most of the chapters have offered examples of good practice. Many of these lessons are transferable and can be used to inform both present and future practice. Although some of the evidence of good practice is specific to an individual country, region or sector, other aspects can either be generalized or are capable of wider application. Three groups or sets of good practice can be identified:

- in most cases, the issue of how to integrate economic development with environmental management is a matter that should be addressed at the start of the policy-formulation process – to attempt to integrate individual entrenched areas of policy is a much more difficult task than designing a policy system that operates through either a single territorial programme (also see the following point) or through a mechanism for policy development and implementation that attempts to integrate the goals and the means of achieving policy at the outset;

- related to the above point, it would appear from the case-study evidence presented in this book that the integration of economic and environmental policies is easier in systems of policy development and governance that permit a high degree of local and regional discretion and autonomy – the ability to tailor the structure and content of a policy system to the environmental, social and economic characteristics of a locality or region, is likely to provide a firmer foundation for policy integration than the imposition of a top-down dictat;
- in order to achieve the integration of economic and environmental policy, it is essential to establish a clear framework of both organization and action. The case studies contained in this book demonstrate the merits of adopting a corporate planning and programming approach to the design of both public and private-sector policy.

Techniques, methods and procedures

Reflecting the points related to good practice, it is evident that a direct relationship exists between the adequacy or otherwise of the methods employed to achieve policy integration and the operation of the linked policies in practice. Clearly, there are direct relationships between the capability of policy systems that have been designed as part of an integrated package of measures and the capacity to achieve multiple objectives; and between the extent of policy responsibility that is exercised at a given level and the delivery of integration on the ground. These characteristics suggest that both the institutional and spatial dimensions of policy design merit further attention. In particular:

- The utilization of partnership-based methods and procedures would appear to offer the greatest potential for both the initial integration of the objectives of economic and environmental policies, and for the design and application of methods and procedures of implementation that simultaneously pursue both aspects of policy.
- Whilst some methods and procedures are able to deliver integration between the two areas of policy on a one-off basis, greater consistency and efficiency in the design and application of an integrated package of economic and environmental measures can be achieved if a permanent organizational base is established at the outset. Such an organization might take the form of a level of government, a sectoral or territorial partnership, or a voluntary or quasi-private organization.

- Furthermore, it would appear that the quality of the techniques, methods and procedures that are employed is an important factor in establishing confidence in the capability of an integrated policy system. A quality-driven approach also appears to produce benefits that enable meaningful monitoring and policy review to be undertaken and applied, thereby further reinforcing the quality of integration. Such a 'virtuous circle' of quality and integration would also appear to help to ensure greater efficiency in the use of resources.

Territorial governance and management

Even though it is difficult to prove that the integration of economic and environmental concerns can better be achieved at a particular level of government, there is evidence to support the view that the two aspects of policy are more likely to be considered together as linked individual elements of a wider package of policies that has been developed under a system of governance that emphasizes territorial rather than functional integration. This has been evident for some time (Friedmann and Weaver 1979) and can be demonstrated from a number of the cases presented in this book. The general evidence that is available would also appear to support the view that gains in terms of the more effective integration of economic and environmental policy are usually matched by improvements in the utilization of financial resources. What is less clear from the evidence contained herein, is which spatial level of government (or governance) offers the greatest potential for achieving integration and the promotion of efficiency. On this point, whilst the evidence would appear to suggest that it is at sub-national level that it is more likely that policy integration can be successfully presented and achieved, much depends on the assignment of mandates and areas of competence at the various levels of government. However, generally, there would appear to be a coincidence between a high level of territorial integration of policy and the presence of an established system of government (or governance) in small nations and at a regional or sub-regional level (Wiehler and Stumm 1995). This general relationship would appear to hold true in both developed and less developed nations (Cohen 1993).

Furthermore, some of the chapters in this book provide evidence which suggests that the presence of a settled system of government or governance, preferably guaranteed by a written constitution, strengthens the ability of an organization to maintain an integrated portfolio of policies over the long term. This is certainly the case in the USA, although there is also evidence that indicates the influence which can be exerted over state programmes by alterations to federal expenditure allocations. Similar evidence is presented in Chapter 2 in relation to the situation in the various European nations. Here the simple and direct relationship between the allocation of powers and competence and the strength of territorial integration is further complicated by new policy approaches and instruments, including the introduction of a blend of measures that seeks to promote the achievement of economic and environmental objectives through improved business development, research and innovation, community development and human-resource management.

Responsibility for policy formulation and implementation

Many of the new measures and approaches that have been mentioned above are indicative of the adoption of innovative procedural and organizational structures on the one hand, and the introduction of more effective governmental structures on the other hand. In addition, a number of new approaches to the design and implementation of operational practices can be discerned, including the establishment of partnerships, spatial coalitions and forums, joint strategy and service arrangements, community-based initiatives, sector-specific organizations and special trans-boundary procedures.

Many of these new approaches are either based on, or aim to bring together, the public, private and voluntary sectors. These new methods and systems of organization reflect the growing necessity of pooling powers and resources in order to achieve common objectives. In part these more extensive collaborative arrangements are indicative of the increased severity and frequency of the financial-resource shortages that have emerged in recent decades, but they also represent a recognition that the problems and potentials associated with the integration of economic and environmental policy are frequently more complex and interwoven than was the case in previous eras.

Specific examples of the new institutional and organizational structures have been presented in a number of the chapters, including:

- The formation of regional and local coalitions that are able to command and deploy resources and to direct policies – examples can be identified from Greater Manchester and in the material presented in Chapter 5.
- The establishment of European Union and national frameworks that encourage and support the formation of multi-sector collaborative arrangements in order to deal with the integration of economic and environmental concerns in relation to specific topics such as transport, countryside management, waste management, housing and urban development.
- There is also evidence of some adverse responses to these new approaches – local and regional governments may fear that their powers and resources will be restricted, some private-sector organizations have expressed the view that they are now expected to pay for the provision of services previously provided by the public sector, whilst the voluntary sector increasingly complains that it is left to 'pick up the pieces' in those areas of activity and policy that are either ignored or abandoned by the public and private sectors.
- In addition to the above points, and cutting across individual examples and procedures, it is essential to acknowledge that the arrangement of responsibilities in the policy landscape is much more fluid than it was in the past. The pace and scale of change in the constitutional and organizational arrangements for the management and governance of the relationship

between economic and environmental matters varies considerably between and within countries. However, a common feature that is observable across most policy regimes, is that the pace of change is more likely to quicken than slacken.

Global debates and influences

Many of the points made in the preceding sections reflect the national, regional and local responses that have been made to international agreements and agendas. At one level of analysis, a cascade of policies and policy integration can be identified from the global to local levels. However, this is too crude a model to explain the complexities that are observable in the current policy system. A more sophisticated model is one which considers both the 'top-down' and 'bottom-up' forces and sources of influence, with the result that policy integration is more likely to be successful at a meso-level. Helpful sources of explanation can be found in Stöhr's work on local development (1990) and in the ideas promoted by Camagni that argue in favour of establishing what he refers to as 'overall territorial balance' (1998: 2).

Despite the common assumption that 'top-down' forces dominate local and regional responses, there are also indications of a gradual realization that most global and transnational policy objectives cannot be implemented without the active agreement and participation of subordinate levels of government and governance. This has led to the promotion of greater flexibility and autonomy, and this, together with the new models of organization and partnership that were discussed in the previous section, has enabled a two-way interchange of policy innovation to be established. This can be seen in different regimes, for example, through the promotion by the European Union of 'pilot projects' that provide evidence and illustrations of good practice that can be used in order to both inform practice elsewhere and assist in the review of policy at European level.

Explanation and the adequacy of theory

What is evident from many of the foregoing sections of this chapter is that, whilst it is possible to observe and describe the operation and evolution of economic and environmental policies and their integration across a range of jurisdictions and regimes, much of this observation is random and lacks any real focus. As was argued in Chapter 1, ecological modernization, whilst offering a framework of analysis and theory, is still generally considered to be limited in terms of both its scope and its explanatory capacity. However, as we observed in Chapter 1, the adoption of an ecological modernization perspective does provide some purchase for environmentalists in mainstream economic debate and, on the evidence presented in this book, this purchase is increasing. The importance of ecological modernization is not related to its ability to offer a generic theory that can provide an explanation for the stability of the environment–economy interaction, but in its capacity to provide a basis of procedural theory that can assist in understanding the processes of implementing some of the inherent goals and objectives of sustainable development.

Accepting these limitations to the contribution of ecological modernization both as a theory and as a mode of explanation, there is, nevertheless, considerable evidence that it has offered a robust framework for analysis across a number of different policy regimes. In particular, ecological modernization provides a basis for understanding and comparing the progress of institutional and attitudinal change across various sectors of activity in different places. It is, for example, possible to compare the evolution of different activities by reference to a common scale for measuring progress. In Hajer's (1996) terminology this pathway of progress starts with institutional learning and then moves through ecological modernization as a technocratic project, eventually ushering in a new culture of politics. This new cultural politics would no longer place emphasis on the protection of nature, rather it would 'focus on the choice of what sort of nature and society we want' (Hajer 1996: 259).

In addition to these over-arching theoretical frameworks that attempt to explain and provide an understanding of the nature and inherent instability of the relationship between the economy and the environment, there are a number of other areas of organizational and procedural theory that are worthy of further mention. In particular, it is essential to acknowledge the importance of those aspects of theory that attempt to identify the key elements of the interaction between processes of economic accumulation and control on the one hand, and organization and institutional structures on the other. Whilst the mode of regulation that is used to mediate between legitimate competing interests will vary according to the socio-cultural inheritance of an individual place and the prevailing economic–environmental balance of power, there are also a number of common factors that are evident in these relationships. Although the mode of regulation is not predetermined by either what has gone before or by a single model of behaviour, it is clear that the new institutional relationships that are brought about by sustainable development place particular emphasis on the inseparable nature of production and consumption (Marsden *et al.* 1993).

Furthermore, these new institutional and organizational relationships bring with them the need for the introduction of new process and procedures for the design and implementation of policy. A number of the case-study chapters in the present text offer illustrations of this point, including the material presented by Taylor, Gibbs and Longhurst, Carter and Winterflood, and Patton and Worthington. New procedures, processes and methods of brokering and managing partnerships are essential features of the search for ways of promoting the closer integration of environment and economy.

Finally, at least with regard to this point, it is important to recognize and accept that the search for a single theoretical explanation of the nature of, and processes associated with, the integration of environment and economy is unlikely to prove successful. Because of the complex and dynamic nature of the relationship between the environment and the economy, the issues and perspectives involved are also subject to constant change. This, almost more than anything else, makes it unlikely that a single theory or set of theories will prove sufficient to capture and explain all of the interactions and aspects of the relationship that have been explored in this book.

Looking forward

Moving beyond the analysis of past successes and failures, and of the attempts that have been made to explore and explain the dynamics of the relationship between the economy and the environment, what lessons can be identified that may help to inform future policy? From the evidence presented herein, it is possible to point to three aspects that are of particular importance: first, the desirability of providing an international or national framework of agreement within which a regional or local programme of integrated action can be developed; second, the desirability of ensuring that sufficient institutional capacity, political power and financial capability exists at the level chosen as the most suitable to promote and implement an integrated package of economic and environmental policy; and third, the need to promote the integration of the two aspects of policy at the outset of the policy-formulation process – to attempt to link one policy to the other at a later stage is to invite difficulties and delays. Each of these aspects is now discussed in greater detail.

The first aspect – the desirability of providing an international or national framework of agreement upon which to base regional and local policy – has been explored at length both in this book and elsewhere in the literature. As we argued in Chapter 1, many failures to integrate economic and environmental concerns stem from the excessive and frequently unnecessary compartmentalization of policy areas. By introducing a requirement for the interchange of agendas at the outset of the policy-development process, it is possible to direct both economic and environment policies towards the attainment of a common set of goals. This approach to the construction and implementation of integrated policy or, to put it in terms that are used by the European Union, the closer integration of the 'horizontal' and 'vertical' elements of policy, can be seen in the recent proposals made by the European Commission for the reform of the Structural Funds (CEC 1997). In future, the regional programmes that are supported by the Structural Funds will be expected to deliver both the traditional 'products' of jobs and GDP growth and a number of new 'products'. These new 'products' will include a variety of environmental and social-development objectives. The intention behind the adoption of this 'twin-track' approach is to improve the efficiency and effectiveness of economic-development policy itself, and also to achieve certain environmental and social objectives through the more extensive deployment of existing financial resources.

Above and beyond these specific examples, it can be seen that the present level of policy integration, despite the many remaining failings, represents a consider-able improvement on the situation that obtained even a decade ago. This improvement in the relative performance of policies that are designed to foster and encourage the closer integration of environment and economy concerns is, in part, a reflection of the greater level of international discussion and agreement on such matters that has taken place since the mid-1980s. Without the foundations which were laid at Rio de Janeiro and other conferences, it is unlikely that national governments would have acted in the ways in which they have. A similar process of policy improvement in individual member states can be seen to have been stimulated by the European Union's Fifth Action Programme on the Environment.

The second point – ensuring that an appropriate form of organizational and operational capacity is present at the most suitable level for the integration of policy – builds upon the first aspect. As has been argued in a number of chapters of this book, there is an increasing agreement that whilst it is essential to build upon the foundations established at international and national levels, the detailed development and implementation of an integrated package of environment and economy policies is likely to be more successful at regional or local level. This is, in part, a reflection of the considerable complexity of policy, but it also represents the increasing reality of having to bring together a wide and varied coalition of actors in order to agree the details of the policy agenda. It is less likely now than it was in the past that a single organization or actor can perform all of the required roles and provide all the necessary resources; this is certainly true of big central governments in hard times.

Given the increasing prominence of local and regional actions designed to promote the integration of environment and economy policies, what role is to be performed by each of the key sectors? Once again, it is important to emphasise the essential characteristics of the new mode of partnership working that has emerged in recent years and which can be seen at both the regional and local levels (Roberts and Lloyd 1999). This new model places particular emphasis on belonging rather than leading, and stresses the common factors that are associated with individual places, rather than assigning all problems to an amorphous global realm. Partnerships for territorial quality place considerable importance on the bringing together of what are frequently disparate elements of policy; this can be seen in experiments such as the environment-city movement and in recent UK Government initiatives to build sustainable communities (DETR 1998).

Furthermore, the establishment of effective administration and governance arrangements at a regional or local level can help to reduce the cost of designing and delivering integrated policy. This greater cost-effectiveness can be achieved through the operation of a single territorial programme that brings together all of the various elements of policy in an area-based package that is managed by a single bureaucracy.

This introduces the third aspect for future attention and development – the desirability of stressing the need for integration at the outset of the policy-development process. Much of what has been presented in this book leads to this conclusion, and it is now evident that attempting to retrofit an integrated framework of policy is less effective and more costly than designing and implementing such at framework at the start.

The benefits of integration can be seen to emerge in many guises: as a means of enhancing competitiveness (see, e.g. Barker and Kohler 1997); as a boost to the stimulation (through research and development) and introduction of new technologies and industries (see, e.g. Office of Science and Technology 1998); as a generator of greater social and community involvement in a range of environmental and economic issues (see, e.g. DETR 1998) and as a means of ensuring the protection of endangered habitats and the best use of scarce resources (see, e.g. LGMB 1996). In short, what can be achieved, by integrating the two strands of policy at the outset, is the more efficient and effective use of natural, financial and human resources.

Bringing these three aspects of the current and future relationship between the economy and the environment together in a single programme of policy development is a difficult task, but it is one which could bring about simultaneous improvements in the performance of the economy, the state of the environment and the quality of life. This multiple-objective approach is one which has been commended by the European Parliament. The parliament has argued that it is essential to promote the greater integration of economic and environmental concerns in order to prevent the continuation of present trends which demonstrate that the EU economies are not developing in a sustainable manner because they are still 'characterized by the under-use of labour resources and over-use of environmental resources' (European Parliament 1998: 4). This is the challenge of the future and it is one which will place increased emphasis on the integration of policies for economic development and environmental management.

References

Barker, T. and Kohler, J. (1997) *Environmental Policy and Competitiveness*, Dublin: Environmental Institute.

Camagni, R. (1998) 'Global challenges and local policy responses in the European Urban System', paper given at the 8th Conference on Urban and Regional Research, Madrid, June.

Cohen, M. (1993) 'Megacities and the environment', *Finance and Development* 30(2) 44–7.

CEC (Commission of the European Communities) (1997) *Agenda 2000*, Luxembourg: Office for Official Publications of the European Communities.

DETR (Department of the Environment, Transport and the Regions) (1998) *Opportunities for Change*, London: DETR.

European Parliament (1998) *Sustainable Development: A Key Principle for European Regional Development*, Luxembourg: European Parliament.

Friedmann, J. and Weaver, C. (1979) *Territory and Function*, London: Edward Arnold.

Gibbs, D. (1996) 'Integrating sustainable development and economic restructuring: A role for regulation theory?', *Geoforum* 27(1), 1–10.

Hajer, M. (1996) 'Ecological modernisation as cultural politics' in S. Laska, B. Szerszynski and B. Wynne (eds) *Risk, Environment and Modernity*, London: Sage.

LGMB (Local Government Management Board) (1996) *Using the EU Structural Funds to Advance Sustainable Development*, Luton: LGMB.

Marsden, T., Murdoch, J., Lowe, P., Munton, R. and Flynn, A. (1993) *Constructing the Countryside*, London: UCL Press.

Ministers Responsible for Spatial Planning (1998) *The European Spatial Development Perspective*, Glasgow: Ministers Responsible for Spatial Planning.

Office of Science and Technology (1998) *Foresight: National Resources and Environment – Sustainable Technologies for a Cleaner World*, London: Office of Science and Technology.

Roberts, P. (1995) *Environmentally Sustainable Business: A Local and Regional Perspective*, London: Paul Chapman.

Roberts, P. and Lloyd, G. (1999) 'Institutional aspects of regional planning management and development', *Environment and Planning B* 26(4), 517–531.

Spaargaren, G. and Mol, A. (1991) *Sociology, Environment and Modernity: Ecological Modernisation as a Theory of Social Change*, Wageningen: LUW.

Stöhr, W. (1990) *Global Challenge and Local Response*, London: Mansell.

Wiehler, F. and Stumm, T. (1995) 'The power of regional and local authorities and their role in the European Union', *European Planning Studies* 3(2), 227–50.

Woodgate, G. (1997) 'Introduction', in M. Redclift and G. Woodgate (eds) *The International Handbook of Environmental Sociology*, Cheltenham: Edward Elgar.

Index

Environment Committee of the House of
Commons 121, 128
Environment-led strategy 141, 150, 156
objectives of 148–9
Environmental accounts 193
Environmental Action Programmes (EAPs)
26–7, 71, 86, 117–22 *passim*, 127–8, 156,
206–7, 242, 267
Environmental assets 139, 142–5, 149
Environmental capacity 156–7, 185–8,
192–3, 197
Environmental fora 95, 136–9, 149–58, 172,
247–8, 252; *see also* Lancashire
Environment Forum
Environmental impact assessment (EIA) 30,
34, 44, 96, 115–17, 178, 220
Environmental improvement programmes
29–30, 91, 99, 168–70
Environmental limits 104
Environmental management 33–4, 77; at
conurbation level 199–200; industrial
195–7; regional 198; systems for 214–15;
techniques of 140–41
Environmental Protection Agency, US
46–7
Environmental space 193
Environmental stock criteria 179–85
Ethnic-minority businesses 173
European Commission 26–36 *passim*,
116–22 *passim*, 128, 155, 266
European Court of Auditors 31, 116
European Environment Agency 155
European Parliament 268
European Spatial Development Perspective
259
European Union 7, 115, 169, 264;
development of concern for the
environment 26–8; environmental policies
of member states 28–30; regional
development strategies 30–36, 71, 81; *see
also more specific references*
Expert Group on the Urban Environment,
EU 213
Exploitation of natural resources 40–41

'Factor four' arguments 195
Far North District (New Zealand) 62
Fife 205–21, 259
structure plan 208–9
Fife Enterprise 211
Fifth Environmental Action Programme
(5EAP) 27, 71, 86, 118, 120, 122, 127–8,
156, 207, 242, 267

Financial management in local authorities
109, 152–3
Finland 25, 32–5 *passim*
Fiscal policies *see* taxation
'Footprints' 74, 77, 79, 193
Foreign Office 73
Forrester, J. 233
Forum for a Better Leicestershire 172
France 25, 29–35 *passim*
Friends of the Earth 33, 230, 248
Fyrstad programme 33, 35

Galbraith, John Kenneth 41
Geddes, M. 250
Geddes, P. 7
Germany 25, 29, 32, 215
Giddens, A. 4–5
Glasser, H. 103
Glenrothes 205, 217
Global capacity 193
Global Environmental Challenge Programme
88
Global sustainability criteria 182
Globalization 115, 195
Gloucestershire County Council 240, 243,
248, 251–2
Government failure 3–4
Government offices for English regions 123,
160
Grassroots environmentalism 45–6, 50
Greater Manchester 155, 190, 195–201, 263
Greater Manchester, Lancashire and Cheshire
(GMLC) project 123
Greece 30
Green audit 227–31, 235
Greenfield sites 151, 156, 188
'Greening'; environmental 208; of economic
development 88, 90, 129, 200–201, 232,
251
Gross domestic product 28
Growth poles 200
Guidance; on environmental matters 119,
121, 128; *see also* Planning policy
guidance; Regional planning guidance

Hajer, M. 265
Hardin, G. 103
Hardy, D. 226
Harmonization within the EU 115–16
Hastings (Nebraska) 47–8
Healey, P. 238
Health risks 43
Heavy goods vehicle (HGV) traffic 175

Milton Keynes UK
Ingram Content Group UK Ltd.
UKHW040445071024
449327UK00020B/1011